UNIPA Springer Series

The **UNIPA Springer Series** publishes single and co-authored thematic collected volumes, monographs, handbooks and advanced textbooks on specific issues of particular relevance in six core scientific areas. The issues may be interdisciplinary or within one specific area of interest. Manuscripts are invited for publication in the following fields of study:

1- Clinical Medicine;
2- Biomedical and Life Sciences;
3- Engineering and Physical Sciences;
4- Mathematics, Statistics and Computer Science;
5- Business, Economics and Law;
6- Human, Behavioral and Social Sciences.

Manuscripts submitted to the series are peer reviewed for scientific rigor followed by the usual Springer standards of editing, production, marketing and distribution. The series will allow authors to showcase their research within the context of a dynamic multidisciplinary platform. The series is open to academics from the University of Palermo but also from other universities around the world. Both scientific and teaching contributions are welcome in this series. The editorial products are addressed to researchers and students and will be published in the English language.

The volumes of the series are single-blind peer-reviewed.

Book proposals can be submitted to the *UNIPA Springer Series Technical Secretariat* at unipaspringer@unipa.it
At the following link, you can find some specific information about the submission procedure to the Editorial Board: https://www.unipa.it/strutture/springer/

Laura Inzerillo

The New Horizons of Oblique Axonometry

The Geometry of the Drawing Geometric Experimentation

UNIVERSITÀ
DEGLI STUDI
DI PALERMO

Springer

Laura Inzerillo
Dipartimento di Ingegneria
Università di Palermo
Palermo, Italy

ISSN 2366-7516 ISSN 2366-7524 (electronic)
UNIPA Springer Series
ISBN 978-3-032-07474-4 ISBN 978-3-032-07475-1 (eBook)
https://doi.org/10.1007/978-3-032-07475-1

This Springer imprint is published by the registered company Springer Nature Switzerland AG
The registered company address is: Gewerbestrasse 11, 6330 Cham, Switzerland

To my dad

Foreword

The crux of the matter.

Laura Inzerillo inherits a demanding tradition: that of her father, Michele Inzerillo, whom she herself remembers in a dedication as essential as it is full of affection. Therefore, to appreciate this book as much as it deserves to be understood and appreciated, it is necessary to refer to that tradition and that way of doing research that was typical of Michele and that he himself declares in a short essay of 2007.

The soul of research is curiosity; the managerial methodology of curiosity charts its course, and the vectorial harnessing of curiosity seals its outcome. Without this vital resource, true research does not exist, and for me, lacking it is akin to reaching the absolute zero of thought. Yet, it is a quality with many shadows it can fall ill with morbidity and arrogance, serve as a screen for bias and ambushes, shake fragile balances, and entangle us in a myriad of doubts, splinters of shattered convictions, clouds of fleeting sparks from ephemeral fires.

Research is a challenge to the unconscious, an open door to the horizon, the first stone of the mysterious and undefined arch of knowledge, barely settled into its bed and exposed to the winds of its possible purposes not easily planned, not all foreseeable.

Let us reflect, let ourselves be carried by the words of thought, and perhaps we will find that research is a symptom, an ecstasy of awareness of knowledge, which must be charged with enthusiasm. The enthusiasm of the first time is fundamental both timid and bold the thrill of being instrumental to something greater than ourselves, which, from the hypothesis of open-ended goals, promotes and promises the unparalleled pleasure of producing original and personal knowledge, a gem of the common heritage.

Enthusiasm sometimes clashes with a disordered, conditioned, and conditioning prior mastery, within its apparent limits; it must be supported through a joyful manipulation of reserves, acquired, and often not deeply rooted. The instinctive and lucid revisiting of the indecipherable baggage of knowledge almost never ready to support the intuitive moment can reveal the path of good fortune, capable of rewarding the intensity, liveliness, and persistence of inquisitive fervor, with the intimate hope of triumph. A mention was made of the non-essential relationship between enthusiasm and the support of solid foundations of knowledge, organically linked to the values of the unknowns at play.

The success of research may depend precisely on the richness of reflection, the appropriateness of considerations, the refusal to think with someone else's head, the doubt of dogmatism, and the willingness to reexamine it.

Therefore, it should not surprise or mislead that the voice of this book by Laura sounds so out of the chorus.

Not only does this book deal with an obsolete subject, but it does so in a way that has now been almost completely abandoned, namely by means of synthetic geometry. Or, as Piero Della Francesca would say, with 'the strength of lines'.

Obviously, the fact that the drawings are made by means of a computer does not matter at all: it is a tool like any other, certainly more accurate, but less fascinating than the burin of the engravers who illustrated the books of the nineteenth century.

So, given that there are no elaborations of three-dimensional models, nor renderings, nor other 'special effects', what is the novelty of this work?

Essentially, I would say, in the treatment of oblique axonometry as a method of representation.

Now, as is well known, axonometry is usually practiced by resorting to construction by points. Thus, the image of an object is obtained by constructing the projections of a certain number of points that describe it, one at a time, by means of the constructor segment, that is, of three contiguous segments, parallel to the coordinate axes in a Cartesian system, each of which is as long as the coordinates of the point itself prescribe. Therefore, a repetitive, tedious, and ultimately obtuse procedure, in which no account is taken of the properties of the object represented, for example, the angular relationships between lines and planes, and the lengths of segments that are not parallel to the reference axes.

Treating axonometry in this way, therefore, means freeing it from all the limits of conventional treatment, and it is not easy. Others have done so, but Laura keeps a completely original way.

And here I come to the heart of the matter: cui prodest?

I do not want to evade this question, because it is the objection to which our daily approach to research, which is technological and pragmatic, to put it mildly, inevitably leads us.

I would like to respond by saying that intellectual efforts, such as the one that appears to have been made here, serve culture and science more than repeated applications of the latest electronic invention. They serve to keep alive the relationship with the history of science that we cultivate, not slavishly repeating its results, but rediscussing them, criticizing them, overcoming them, as Michele invited us to do in the poetic text I quoted.

Rome, Italy — Riccardo Migliari

Preface

The act of drawing is not merely a means of representation—it is a mode of reasoning, an intellectual process that transforms abstract thought into visible structure. In the field of Descriptive Geometry, this process is particularly evident, and oblique axonometry, though often marginalized in contemporary practice, offers a powerful and versatile instrument within this framework. This book was born from the desire to reassert its relevance, to explore its depths, and to propose a methodical, clear, and pedagogically valuable approach to its study and application.

Over the last few decades, the evolution of drawing techniques has been increasingly dominated by digital tools, 3D modeling, and photorealistic rendering. While these tools undoubtedly provide extraordinary opportunities for visualization, they sometimes risk distancing the designer or student from the geometric reasoning that underpins spatial understanding. Oblique axonometry, with its synthetic logic and representational autonomy, invites a return to an essential discipline—one that does not seek to imitate reality but to construct it through precise, conceptual tools.

This volume addresses both beginners and advanced scholars, offering a gradual and layered progression through the method. It begins with the theoretical principles of oblique projection, proceeds through the reconstruction of coordinate systems, and culminates in the interpretation of complex geometries: polyhedra, curved surfaces, and ruled or helical forms. Each chapter combines rigorous theoretical exposition with clear operational strategies, supported by geometric constructions that can be replicated, tested, and developed independently.

The goal is twofold: to provide a structured and comprehensive didactic tool, and to stimulate a reflective attitude toward geometric drawing. Rather than offering mere technical instructions, the text invites the reader to question assumptions, recognize the underlying rules of projection, and understand how form behaves when abstracted onto the drawing plane. Geometry here is not reduced to a set of mechanical rules, but is presented as a living language—one that speaks of space, logic, and invention.

Oblique axonometry, in particular, resists simplification. It requires from the drawer an active involvement: to choose the angle of projection, to interpret distortions, and to define relations among axes and planes. These challenges, far from being limitations, are opportunities. They foster awareness, sharpen critical thinking, and demand a constant balance between intuition and rigor.

This work also pays homage to a tradition of teaching and research in representation sciences that values synthesis over simulation, clarity over spectacle, and deep understanding over superficial reproduction. It is a contribution to the revalorization of geometry not only as a technical foundation but as a cultural and expressive tool—a heritage to be cultivated and renewed.

I hope that this book will inspire readers to draw not only with precision, but with curiosity. To explore space through reasoning. To rediscover the pleasure of constructing rather than imitating. And to recognize, once again, the silent but eloquent force of the line.

Experimentation in the field of Descriptive Geometry, a fundamental component of drawing disciplines, constantly generates new stimuli for theoretical and practical investigation. This monograph explores one of the most underappreciated yet profoundly versatile techniques: oblique axonometry.

While recent developments in geometric representation have often focused on automation, digitization, and photorealistic reconstruction, this work chooses to return to the core of geometry—its synthetic strength and intellectual rigor. The pages that follow aim to offer a fresh, structured, and methodologically sound approach to oblique axonometric projection, including both its theoretical underpinnings and a wide range of spatial constructions.

The chapters are structured progressively: from the foundational principles of oblique projection, through the geometric reconstruction of reference systems, to the reading of planes, lines, and complex volumes such as ruled, toroidal, and helical surfaces. Each chapter is designed to balance precision and accessibility, with the intention of empowering the reader to apply these methods in both academic and professional contexts.

This work aspires not only to reaffirm the pedagogical and expressive value of synthetic geometry but also to provide inspiration for further research. I hope it will encourage readers to question conventional approaches and to embrace the freedom and depth offered by geometric reasoning.

Enjoy the reading.

Palermo, Italy Laura Inzerillo

Contents

1 Oblique Parallel Projection 1
Image Constructed According to the Geometry of the Drawing 1
Geometry of Space and Geometric Projection System 3

2 Geometry of the Drawing in Oblique Axonometry 5
The Triplet O′, x′, y′, z′, Oblique Projection from V onto π of the Triplet O, x, y, z 5
Restitution of the Reference System and Obliquity of Projection 7
Remarkable Trio Choices 9

3 Oblique Axonometry: Coordinated Plane Parallel to π 11
Representation and Reading of Points, Lines, and Planes 11
Projection or Restitution of a Plane Figure 22
Squareness and Distances 26
Translation and Rotation 35
Central, Axial, Orthogonal, and Oblique Symmetry; Symmetry with Respect to a Plane 42

4 Volumes in Oblique Axonometry 53
Polyhedra 53
Regular Polyhedra 70
Semi-regular Polyhedra 77
Parabolic Point Quadrics 82
Elliptical Point Quadrics 87

Appendix: Research Future Directions 99

Bibliography 101

Chapter 1
Oblique Parallel Projection

Abstract This chapter introduces the foundational principles of oblique axonometric projection, establishing its relationship to orthogonal axonometry while emphasizing its autonomy as a complete system of representation. The theoretical framework is developed through spatial reasoning, without reliance on imagery, guiding the reader to a deeper conceptual understanding of projection geometry. Definitions of spatial elements and their infinite and relative properties are explored, preparing the reader for the practical and intellectual challenges of oblique drawing. The aim is to develop conscious, methodical control of geometric representation, overcoming intuitive but imprecise approaches to descriptive geometry.

Image Constructed According to the Geometry of the Drawing

The projective system under examination, Oblique Axonometry, is a generalization of the one already widely treated, Orthogonal Axonometry; it is not necessary to pre-commit to the study of the latter; The conquest of the new must be set up independently, autonomously, completely.

It can be said that Oblique Axonometry is all-encompassing of the other, which can be found in its entirety, imposing on the oblique angle, between the direction of the projection ray from the center to infinity and the picture, the particular value of the right angle.

The goal of the scholar is to know how to build an image in oblique axonometry, guided by an adequate mastery consciously matured through learning, in-depth study, and experimentation.

If he wants to be such, he must free himself from numerous and historical conditioning preconceptions; he must orient himself on essentially applicative activities, with freedom and wide-ranging.

If this is his action, he will be able to significantly increase the wealth of his theoretical and practical geometry of drawing operational resources.

L. Inzerillo, *The New Horizons of Oblique Axonometry*, UNIPA Springer Series,
https://doi.org/10.1007/978-3-032-07475-1_1

By entrusting his growth to reasoning, considerations, and reflections, typical of the Science of Representation, he will be able to follow new paths, aimed at freeing him from the risk of having to curb his imagination, of having to mortify his creativity, of having to contain his curiosity.

His figurative protagonism must ride the "good" doubt; it precedes and promotes the will to seek the irrefutable explanation, the reassuring resolution; the need for "why" must invariably raise a priori doubts, certainly constructive.

Being or becoming such, in a normally unforeseen confrontation, makes one exempt from the devastating panic of the recurring "bad" doubt; self-esteem comes out in pieces; the heavy awareness of possessing limited competence, an unreliable readiness, an embarrassing and damaging unwillingness to face representative and interpretative questions arises.

The risk of missing or missing motivations for doing should not be underestimated; It is not always and not when one hopes, that it is possible to avoid a report placed or proposed to one's own attention and that of others, in the professional activity or in the process of scientific research applied to geometric space.

The cornerstones of the first impact with the geometry of drawing recall the concepts of Elementary Geometry, on which, in truth, little or no attention is dwelt; The common sense of mnemonic sufficiency is usually a source of bitterness and trouble.

The list of terms conceptually linked to the geometry of drawing is long, unlimited, open; it is unimaginable; We would like to reflect on the most frequently present in graphic applications, with ample reserve to suggest others:

- dimensionlessness, identifiability, site, direction, position, absolute, relative, infinite, finite, indefinite, definite, all, nothing, always, never, uniqueness, multiplicity, exclusion, inclusion, involvement, superimposition, adherence, alignment, center, extreme, reference, belonging, appearance, exposure, apparent contour, true form, size, distance, correspondence, biunivocality, angle, rotation, cyclicity, pitch, rolling, approach, distance, fullness, void, thickness, simple, composite, articulated, assembled, exploded, aggregate, overhang, cavity, symmetrical, specular, concentricity, coaxiality, contiguity, conjugality, tangency, interior, exterior, section, intersection, coplanarity, element in field, element off-screen, incident elements, crooked elements, two-sided surfaces, the boundary, the single-sided surface, the whole, the partial, parallelism, orthogonality, similarity, equality, equivalence, repetitiveness, eurythmy, proportionality, homology, affinity, homothety, movement, translation, rotation, rototranslation, reversal, genesis, duality, projectivity, perspective, continuity, developability, the visible, the invisible, etc.

The scholar, on each of these terms submitted to his analysis, cannot fail to dwell, reflect, and respond with an abundance of arguments and appropriate examples.

He cannot help but hypothesize other connective terms between thinking and doing; he cannot fail to want to manage a wealth without borders, rampant and yet almost always under control.

In part, it is a wealth of knowledge to be exhumed; evidently, reminiscence is not enough; it is necessary to update to more advanced levels.

In the author's handouts, used for many decades by his students of Civil Engineering and then Environmental Engineering, as recommended reading, you can find some of the considerations.

They are stimuli to "think beyond", to challenge each other, to break the common limits of the program, the common learning goals, often disrespectful of the potential of young people.

Geometrical thought strives to overcome, as the engine of science, the limits between the known and the unknown, but it cannot succeed; at most, it can move the boundary polygonally.

Human thought cannot fail to recognize its insurmountable limits in the face of the dimensionlessness of geometric entities, of the point, of the line, of the plane, of the surface, and of space.

Can the finite and the infinite be compared? The rational and the irrational, the real and the imaginary, the relative and the absolute, the all and the nothing, the never and the always, the commensurable and the immeasurable, the temporary and the eternal, the discrete and the continuous, etc.?

However, the scholar should reflect, evaluate, plan, get back in motion, overcome every dilemma, and with compromise, with hypothesis, with convention, with appearance, with the involvement of the virtual, the imaginary, the absurd, manage the use of practical contributions.

By virtue of all this, one can continue to think, to reason about the geometric sign; It should be emphasized, in the meantime, that we have spoken of the geometry of representation and, however, it has not yet been necessary to draw or exhibit an image of it.

Geometry of Space and Geometric Projection System

In the representation in Oblique Axonometry, one begins to become aware of geometric space, when one thinks of a point V, improper; it is a dimensionless point.

Perhaps, one has the sense of one's own place, but of the dimensionless and the improper, no!

Do not dare to understand it as a point of view; To mistake it for the eye is irrational!

Therefore, think, because we can only think of them, the point at infinity, V, and a plane picture π; assign to V a function entirely and only of its own, geometric: it is the center of projection of all geometric space, globally, in all the meanings and limitations indicated and indicable.

Choosing the direction of an improper point V, at points A, B, …, N, … of a given arrangement of space, the projecting rays V-A, V-B, V-…, V-N, … and, given a plane π of position neither orthogonal nor parallel to the direction V, the aforementioned projecting rays intersect π.

These rays identify the points A′, B′, …, N′, …, which show in projection how the points of the three-dimensional spatial arrangement appear in a two-dimensional figurativeness.

The projection T′ of a point T of π from V to π coincides with T, the two points are joined.

Let P′ be the intersection on π of the projecting ray V-P, in the possible sequences V-P-P′ or V-P′-P.

Let the projecting rays V-P-P′ and V-Q-Q′ be parallel to each other; for V oblique on π, they are equally oblique on π; the point P corresponds to P′, the point Q′ corresponds to the point Q′; if Q belongs to the line V-P, the point Q′ will coincide with P′; from P we pass to P′ and from P′ we cannot identify P.

A triaxial trirectangle reference O, x, y, z can be introduced, so that, in geometric space, each point P can be identified by the trio of coordinates P_x, P_y, P_z, or by P and by one of them of your choice, or by P and P_{xy} or by P_{yz} or by P_{zx}.

If in the oblique axonometric projection from V to π, the aforementioned triplet O, x, y, z, is in some way reconstructable, starting from its projection O′, x′, y′, z′, then P can be relocated to its site linked to the spatial triple, starting from the projections of its projections $P_x{}'$, $P_y{}'$, $P_z{}'$, or starting from its projection P′ and the projection of a second projection, $P_x{}'$ or $P_y{}'$ or $P_z{}'$, or starting from P′ and the projection of a second projection of it, $P_{xy}{}'$ or $P_{yz}{}'$ or $P_{zx}{}'$.

All this must be clear, not through a drawing, but simply by reasoning in space, thinking, answering one's question "sure?" "Yes! Certainly".

Chapter 2
Geometry of the Drawing in Oblique Axonometry

Abstract This chapter delves into the geometric mechanics underlying oblique axonometric projections. It focuses on the projection and restitution of coordinate systems, starting from an oblique projection of the standard triplet (O, x, y, z) onto a picture plane. Through analytical constructions and homological relationships, the chapter illustrates how to determine obliquity, evaluate projections of geometric elements, and calculate distortions. Emphasis is placed on the role of geometric reasoning in defining projection parameters and on how variations in angles and positions affect the resulting image. The chapter lays the groundwork for consistent and reproducible drawing practices in oblique axonometry.

The Triplet O′, x′, y′, z′, Oblique Projection from V onto π of the Triplet O, x, y, z

Let T_x, T_y, T_z, (Fig. 2.1) be the traces of the axes of the triplet O, x, y, z, t_{xy}, t_{yz}, t_{zx}, the traces of the coordinate planes; and t_{xy} horizontal.

Project O, which is thought behind π, orthogonally on π, obtaining O_o', orthocenter of the triangle of traces, and x_o', y_o', z_o', orthogonal projections of x, y, z.

The semi-circumference of diameter T_x-T_y locates (O_o); H_o-O_o'-(O_o) defines the ratio of the orthogonal affine homology of axis t_{xy}, being O_o', (O_o) homologous.

Given on (xy), reversed by xy, a figure, the square (A)-(B)-(C)-(D), the projection A_o'-B_o'-C_o'-C′ is obtained; for example, C′ is obtained either from (D)-1-2-C′, or by relying on the homologous lines (O_o)-3-O_o'.

Invert the projecting plane γ of trace H_o-T_z; we have in H_o-C_o* and $z^* = O_o^*$-T_z the true length of H_o-O_o' and O_o'-T_z, and in O_o'-O_o* the true distance of O from π.

Keep the plane xy, the H_o-O_o*, the line of maximum slope of xy with respect to π overturned on π with γ, (O_o), and H_o-(O_o) unchanged; project O obliquely on π in O′ according to the angle φ, defined in overturned by O_o*-O′.

In the orthogonal projection system, the length H_o-O_o* is H_o-O_o', in the oblique one, it is H_o-O′; there is still an orthogonal affine homology between (xy) and x′y′

L. Inzerillo, *The New Horizons of Oblique Axonometry*, UNIPA Springer Series,
https://doi.org/10.1007/978-3-032-07475-1_2

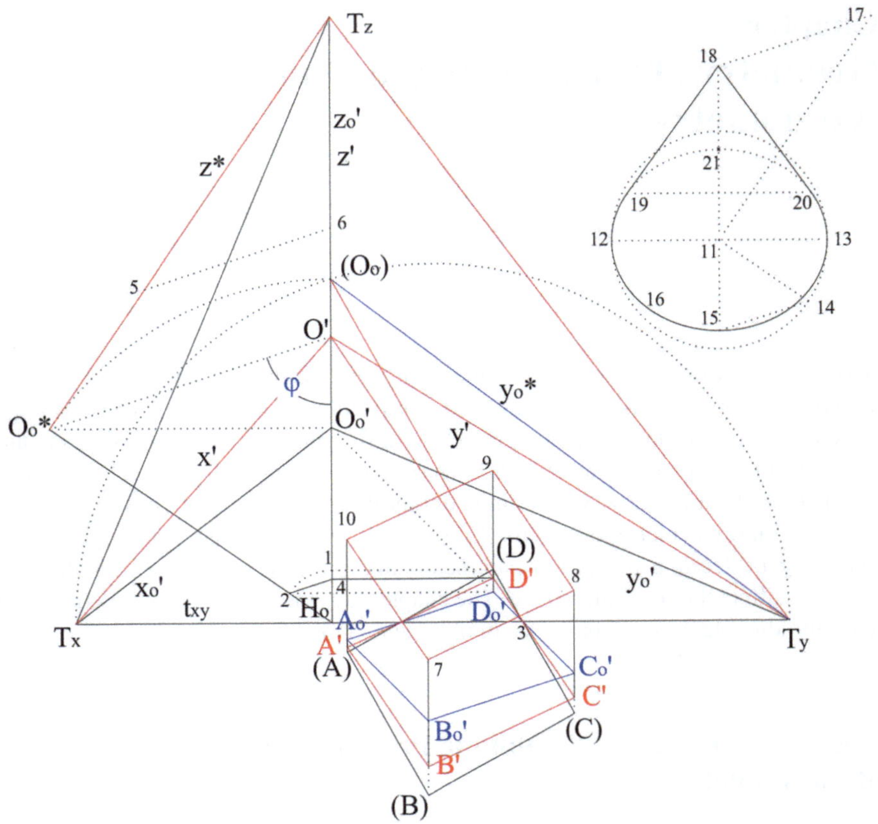

Fig. 2.1 Reference system and projection obliquity: Restitution of the trirectangular system using affine homology

of axis t_{xy} and homologous lines (O_o)-3-O′; squared (A)-(B)-(C)-(D) corresponds to A′-B′-C′-D′; e.g., D′ is obtained from (D)-1-2-4-D′, being 2-4 parallel to O_o*-O′.

If φ changes, O′ moves and what happens can be easily predicted; if O′ reaches O_o′ we have orthogonal axonometry, already known, and A′-B′-C′-D′ coincides with A_o′-B_o′-C_o′-D_o′; if O′ reaches (O_o), A′-B′-C′-D′ coincides with (A)-(B)-(C)-(D).

It is excluded that O′ moves to H_o, because obliquity would involve xy projecting on π, and any of its figures would coincide in oblique projection with the line t_{xy}.

Note that the oblique projection of the figures of xy depends on the homological ratio H_o-O′/H_o-(O_o), i.e. it can be defined by assigning H_o, O′, (O_o); it can be worth 1, less than 1, more than 1, it can be negative; The angle of obliquity ranges from right to zero positive or negative.

The approach of the transition from orthogonal to oblique projection is completed, considering the shortening of the orthogonal dimensions to xy made not according to O_o*-O_o′, but according to O_o*-O′; e.g., if O_o*-5 = (A)-(B) is chosen, it is shortened according to 5-6 in O′-6 and returned to B′-7; This is followed by the projection of the cube resting on XY.

It should be noted that the choice, for convenience, of the angle of obliquity on the vertical projecting plane γ does not invalidate or condition the absolute validity of the previous reasoning.

The variation of the distance O_o*-O_o′, i.e., of O from π, leaves the figurative result of the volume unchanged; note that from x-y derive the distinct pairs x′-y′ and x_o-y_o, while from z derive z_o′ and z′ superimposed.

Segments parallel to t_{xy} retain direction and size; orthogonal ones shorten, or lengthen, like H_o-(O_o) in H_o-O′.

Let the circumference of center 11 and diameter 12-11-13 be given on xy, the major axis of the ellipse 16, the oblique projection of the circumference, and in true magnitude, being parallel to t_{xy}; by means of the 11-14 parallel to the H_o-O_o* and the 14-15 parallel to the O_o*-O′, the semi-minor axis is obtained and the ellipse 16 is constructed.

Once the height of the right circular cone resting on xy is reported in 11-17 parallel to z*, it is shortened to 11-18 according to the 17-18 parallel to O_o*-O′; the tangent apparent contour generators 18-19 and 18-20 complete the image of the cone; The base arc 19-21-20 of the 16th is not in sight.

Restitution of the Reference System and Obliquity of Projection

Let (Fig. 2.2) be given the free projection of a volume, asserting that it is a cube in axonometry; it is possible because the opposite sides are parallel and equal, both for the sides of the lower base A′-B′-C′-D′, and for those of the upper base 1-O′-2-3, and for those that connect the bases.

Consider, on the plane of the upper base xy, the horizontal trace t_{xy}, the traces T_x of x′ = O′-1 and T_y of y′ = O′-2, the semi-circle 4 of diameter T_x-T_y, which intersects D′-O′ in (O_o).

By orthogonal affine homology, the true magnitude (Oo)-5 of the side of the cube is identified; the angle φ of obliquity and O_o* must be obtained such that the shortening of the vertical side equal to (O_o)-5 results in O′-D′.

The arc of center H_o and radius H_o-(O_o) intersects the arc of center (O_o)-5 in O_o*; the horizontal O_o*-O_o′ and the angle φ = O_o*-O′ of obliquity are defined; the comparison with the previous exercise confirms the correctness of the restitution.

Let the projections u_x′ = O′-1 on x′, u_y′ = O′-2 on y′, u_z′ = O′-10 on z′, of the unit u, unique for the three axes of the trirectangle triad of reference, be given (Fig. 2.2b).

By orthogonal affine homology of axis txy relative to the circumference 3 of diameter T_x-T_y, we obtain O_o*, u_x* = (O_o)-5 and u_y* = (Oo)-4, equal to the unit u, the arc 6 of center H_o and radius H_o-(O_o), the arc 7 of center (O_o)-5, their intersection

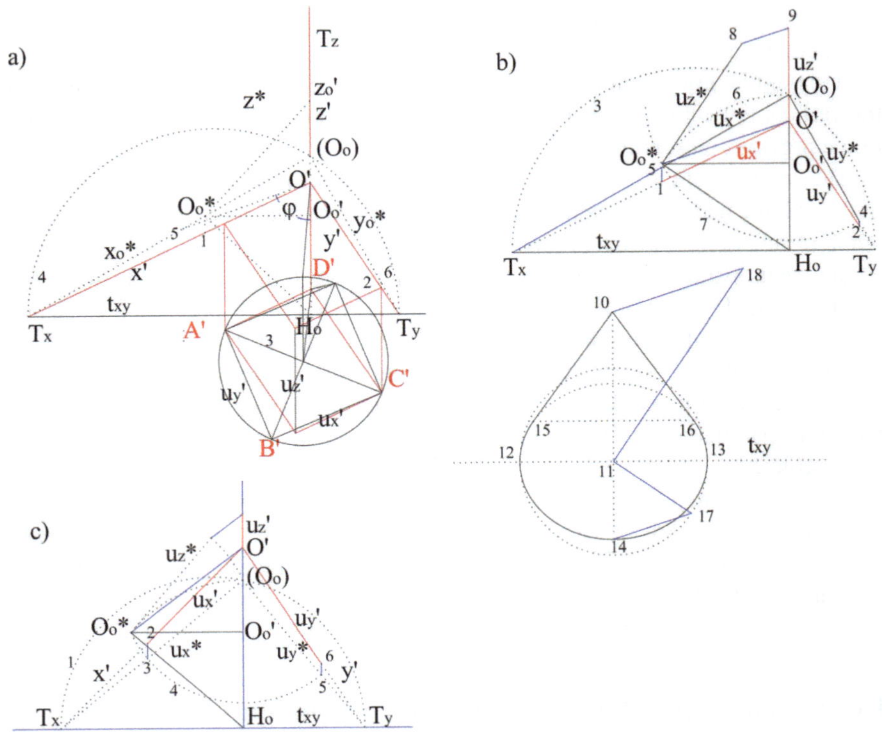

Fig. 2.2 Reference unit measures and angles: Determination of unit vectors and obliquity angle from the projected cube

O_o*, the O_o*-O_o' and the O_o*-O′, the angle of obliquity φ in O′ between O_o*-O′ and O′-H_o, the H_o-O_o*, the z* orthogonal in O_o* to the H_o-O_o*.

From $u_z* = u_x* = 5\text{-}(O_o) = u_y* = u$ reported in O_o*-8, through the 8-9 parallel to O_o*-O′, we find $u_z' = O'\text{-}9$; The elements obtained make it possible to return, for example, the cone, of which the image is known, linked to the assigned triad.

It is verified that the elliptical base is an oblique axonometric projection of a circumference; the radius 11-17 parallel to the H_o-O_o* and the 17-14 parallel O_o*-O′ find the semi-minor axis 11-14.

The value of the height 11-10 is obtained through the 10-18 parallel to the O_o*-O′ and the 11-18 parallel to the O_o*-8; The return is determined by the radius of the base circumference 12-11 and the height 11-18.

The vertex angle O′ between x′ and y′ (Fig. 2.2) can be less than 90°; given O′, x′, y′, z′, and on x′ u_x', having chosen the t_{xy} orthogonal to z′, the traces T_x and T_y, the H_o, the 1, (O_o), x* and y*, from $u_x' = O'\text{-}2$ the $u_x* = (O_o)\text{-}3$, the arc 4 of center (O_o) and radius u_x*, which determines on one side $(O_o)\text{-}5 = u_y*$, from which derives the $O'\text{-}6 = u_y'$ and on the other intersects in O_o* the arc of center Ho and radius H_o-(O_o).

This is followed by H_o-O_o*, O_o*-O_o', O_o*-O′, the reversed direction of obliquity, and for O_o*, orthogonal to H_o-O_o*, the z* on which $u_z* = u_x*$ is reported from O_o*; for the extreme, we consider the parallel to the O_o*-O′, obtaining u_z'.

Therefore, given any triple, without any constraint, and established a value u_x', we can derive the trirectangle triplet in the axonometric space, whose projection, with the consequent obliquity, is precisely that given.

Remarkable Trio Choices

Report, (Fig. 2.3), case (a), the triplet O′, x′, y′, z′ of xy; H_o-O_o*, overturned by the line of maximum slope on π, forms the angle a with π; a point of xy such as O, in an inverted position, in the projection of obliquity φ is found in O′, while in the orthogonal projection it would be found in O_o'.

Note that, if the unit (O_o)-1 on x_o*, of oblique projection O′-3, is, by choice, equal to the unit (O_o)-2 on y_o*, of oblique projection O′-4, it results in O′-3 = O′-5

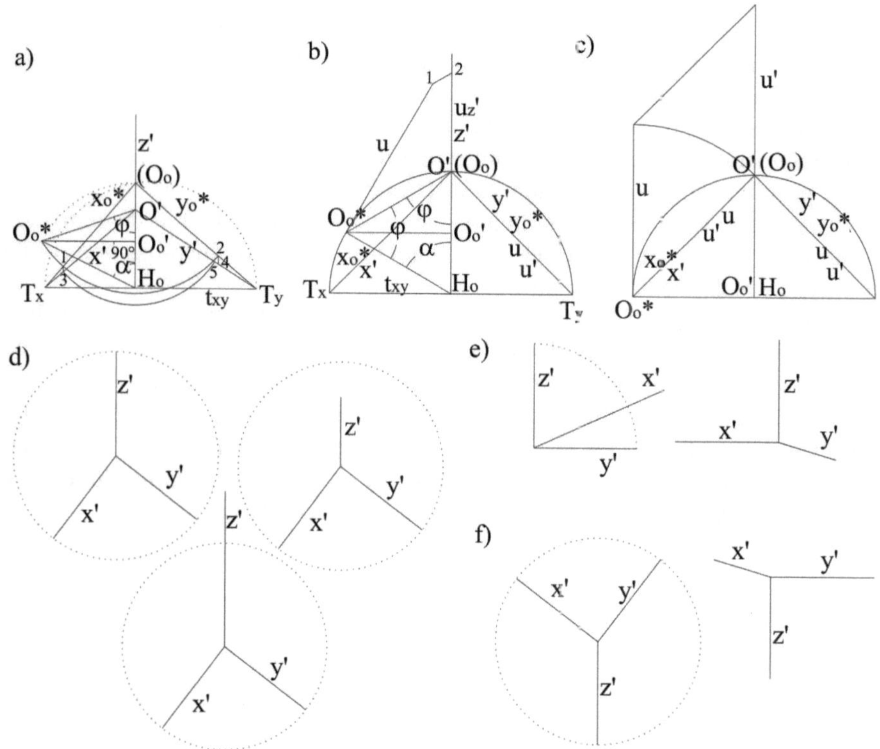

Fig. 2.3 Reference triplet choices: Variations in axis configurations and resulting image transformations

$< O'$-4; if Ho is mean of T_x-T_y, the projections of the two equal units are equal and symmetrical with respect to z'; O_o* is located on the semi-circumference.

The possible variation of α forces us to investigate its effects; predictable results must be able to be derived from hypothetical or hypothetical data.

What happens if α is 45°, 60°, 90°, 0°, if the angle H_o-O_o*-$O' = H_o$-O_o*-$O' = j$, if φ varies, if O' falls outside the range H_o-(O_o)? Reflect, do not give up, it is essential to approach what is affordable, to identify notable cases.

In case (b), we set: the angle x'-O'-$y' = 90°$, $\alpha = \varphi = 60°$, $u_x = u_y = u = T_x$-(O_o) $= T_x$-$O' = u'$, 1-2 parallel to T_x-O', $u_z = u$, $u_z{}' < u$; in case (c) $\alpha = 90°$, $\varphi = 45°$, $u_x = u_y = u_z = u_x{}' = u_y{}' = u_z{}' = u = u'$.

If the projection of the right angle between x and y is to be 90°, the figures of xy are projected unchanged, preserving their shape and size (Fig. 2.3d); this also occurs in the case where the xy plane is parallel to the frame π, and if the projection obliquity is 45°, the value $u = u'$ is unique and unchanged.

It should be noted, however, that two segments orthogonal to each other and in any case direct, belonging to xy parallel to the frame π, in projection with an obliquity of 45° retain the angle and length.

If the two orthogonal segments belong to yz or zx and are parallel to the axes, they retain their size, but the angle between the projection's changes; they remain parallel to the projections of the axes, which do not appear orthogonal.

In coordinated planes, defined in projection by non-orthogonal lines, the right angle between directions parallel to the axes, in projection, can be identified as the angle between conjugate directions, different from the right line, except in the case of directions parallel to the axes of the projection ellipse of the circumference belonging to the position under consideration, as will be clarified later.

Depending on the angle of obliquity, the projection of the unit of the third axis, for example, on z', can be greater or less than that relative to the projections of the two axes, which remain orthogonal.

In (e) you have the view from above, while in (f) you offer the view from below; observe the visual effort to identify the aforementioned choice, until the correct reading of the space is stabilized with reasoning.

Note that the three real units are equal; in projection they can differ; in particular, reflect on case (a) for O' superimposed on (O_o); the projections $u_x{}'$, $u_y{}'$, $u_z{}'$ are different; In this case the angles of the figures of XY are preserved, but the dimensions change; The angle α is not zero and XY cannot be parallel to π.

Chapter 3
Oblique Axonometry: Coordinated Plane Parallel to π

Abstract This chapter explores cases in which the projection plane p is parallel to one of the coordinate planes, particularly focusing on xy. It examines the simplifications and visual effects produced by this condition and offers a detailed analysis of the representation and reading of points, lines, and planes. Through geometric constructions, the text presents how to identify traces, intersections, and projections of spatial elements under specific obliquity conditions. Practical exercises guide the reader through the interpretation of drawing distortions and projection-induced transformations. The chapter also highlights the challenges of interpreting spatial data in a flat image.

Representation and Reading of Points, Lines, and Planes

Simplification, as a requirement of representing and reading a geometrically constructed image, involves the choice of a particular approach while preserving the generalized conceptual character.

The hypothesis of the plane of the frame π parallel to one of the coordinate planes of the spatial reference triplet is certainly simplifying; given, for example, xy is parallel or even coinciding with π, it is of no relevance that x' forms a predetermined angle with the horizontal, orthogonal to z'.

Therefore, the classification of different oblique axonometry is considered an "oversight", according to whether x', relative to xy parallel to π, forms an angle of 30° with the horizontal and y' of 60°, or of 45° and y' also of 45° or of 60° and y' of 30°.

Given the triplet O, x, y, z', the point P_i of space is defined by P_{ixy}, P_i', or, if you like, by P_i', P_{ixz}', or by P_i', P_{iyz}'; the point P_i can belong to xy, to xz, to yz, to x, to y, to z, with respect to xy it can have positive or negative altitude, being located in one of the eight regions of space thus divided by the coordinate planes.

Its real coordinates referable to O, x, y, z, at a single graphic scale, are projected along the three axes according to their respective angle with π; depending on their

L. Inzerillo, *The New Horizons of Oblique Axonometry*, UNIPA Springer Series,
https://doi.org/10.1007/978-3-032-07475-1_3

value, they can fall within the graphic field, or be out of range, either as a point or as a projection point on one of the coordinated planes.

An improper point cannot be on the pitch; it is represented by the improper points of the two projections r_i', r_{ixy}', or r_{iyz}', r_{izx}', of a line r_i, which contains it; a point P belongs to the line if two of the projections of P are located on the projections of the same name.

It is not considered to illustrate with special figures the representation of the point, in the generic and particular positions mentioned above, as they are implicit in that of the lines and planes; in reading, the attribution of false belongings should be avoided.

Let the x and y axes of xy belonging to p be assigned and project z into z′ according to the angle of obliquity $\varphi = 45°$, declared by $u_z' = u$; without this indication, it is impossible to read the drawing.

In Fig. 3.1, the projection of the triplet is represented; x and y, orthogonal, are on π; the z axis is orthogonal to π in O, in space; projecting it obliquely at 45° constitutes z′.

The line a projects into a′; a′ is a datum and does not by itself define the projection of a; another datum is needed; for example, the projection of its orthogonal projection onto xy, a_{xy}, without superscript because it is on π.

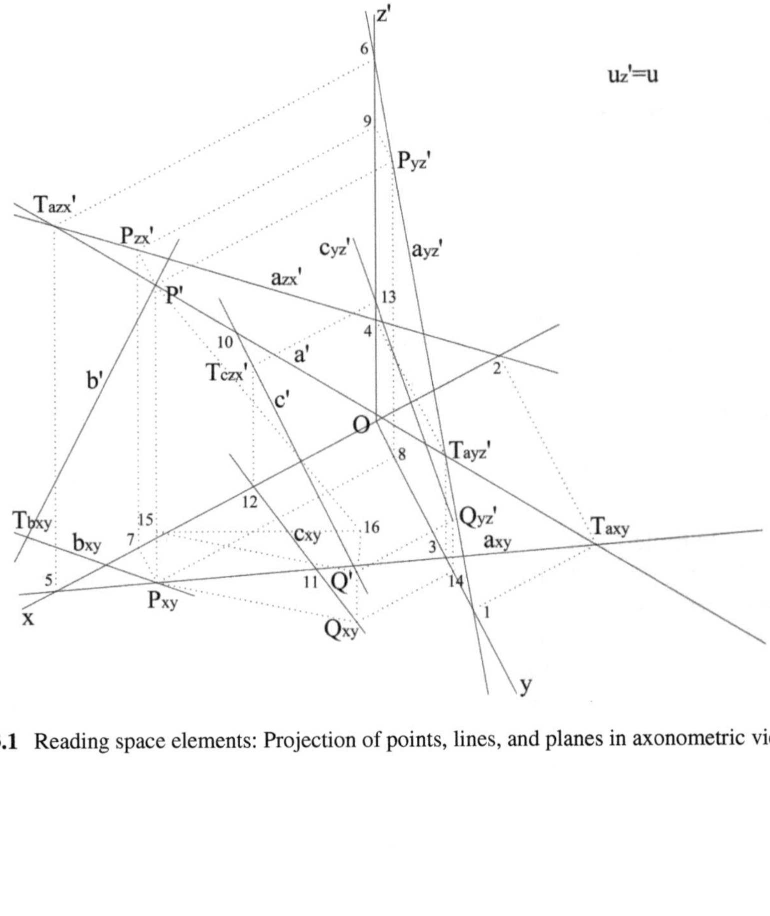

Fig. 3.1 Reading space elements: Projection of points, lines, and planes in axonometric view

The a intersects any plane of space; meets xy in the track T_{axy}, a point common to a′ and a_{xy}; as a point of xy, the T_{axy} has three projections, the one on xy, whose projection is superimposed on the same point, the one on yz, whose projection is 1, on the y-axis, and the one on zx, whose projection is on the x-axis in 2.

The a′ intersects the plane yz in the T_{ayz}' trace, a point common to the a′ and the a_{yz}' projection of the orthogonal projection of the a on yz; the T_{ayz}, as a point of yz, has three projections, on xy, the 3 on the vertical for T_{ayz}', and the 4 obtained on z′ with the parallel to y conducted for T_{ayz}'; the a_{yz}' is obtained using 1 and T_{ayz}'.

Similarly, arguing on the third trace of projection T_{azx}' derived on a′ from 5 or 6, having 2 and 4, we have points in excess to obtain a_{zx}'.

On the generic line a, consider the point P of projection P′; the projections of its orthogonal projections P_{xy}, P_{yz}', P_{zx}' belong to the respective projections of a; P_{xy} is on the parallel to z′ for P′, P_{yz}' is on the parallel to x for P′, P_{zx}' is on the parallel to y for P′; the connection between P_{xy}, P_{yz}', P_{zx}' is made through points 7, 8, 9.

The line b, of which the projections b′ and b_{xy} are given, contains the point P of a; a and b are incidents in P; considering a third line, c, of projections c′ and c_{xy}, the absence of intersection between c and a is observed; c′ intersects the a′ in 10 and the c_{xy} intersects a_{xy} in 11 and between 10 and 11 there is no vertical alignment; The lines a and c are crooked.

It should be noted, again, that T_{czx}' is in the field, while T_{cxy} and T_{cyz}' are off-screen; if we want to obtain the c_{yz}', we choose a point Q of the c, given by Q′ on the c′ and by Q_{xy} on the c_{xy}, we obtain the Q_{yz}' through the 14 on y; from T_{czx}' we arrive at 13 on z′; the 13-Q_{yz}' is the c_{yz}'.

The point P in space is placed at a distance from π equal to P′-P_{xy} on the orthogonal to the drawing sheet, which is π, conducted through P_{xy}; similarly, for Q, we imagine to see it on the orthogonal in Q_{xy} at a distance Q′-Q_{xy}; this is true for $u_z' = u$.

The distance between P and Q is obtained on the plane defined by the two verticals of the distances from π, reversed on π for the rotation around the line traced P_{xy}-Q_{xy}; by bringing the difference in the dimensions orthogonally to the P_{xy}-Q_{xy}, the distance between the two points can be read.

If Q′-Q_{xy} is returned to Pxy-15, the triangle P′-15-Q′ in space is a rectangle and its cathetes P′-15 and 15-Q′ are real; just arrange them at 90°; the hypotenuse P′-16, relative to the cathetes P′-15, difference in the altitudes, and 15-16 = 15-Q′, distance of the feet of the verticals of altitude, both separated in true length, represents the distance between the two points.

The scholar should elaborate on what has been illustrated above, referring the elements considered no longer to xy but to yz or zx; he can go even further, using as a framework π no longer the coordinated plane xy, with orthogonal axes, but the plane yz, with orthogonal axes and x projected obliquely at 45° in x′ indicating $u_x' = u$, That is, the plane zx, with orthogonal axes and y projected obliquely at 45° in Y′ indicating $u_y' = u$.

Considerations and reflections support the scholar to take bolder steps; for example, it can represent the line with a direction that is no longer generic, but particular; there are orthogonal lines to the three coordinated planes; to obtain traces and projections; there are lines parallel to the coordinate planes, the lines belonging to them, the lines incident the axes, the lines for O, the projecting lines, etc.

Of these, certainly, there will be opportunities to illustrate their intuitive peculiarities; a significant gradient of effectively constructive self-esteem can be drawn from the success of the suggested attempt.

Given x-y′-z with $u_y' = u$, (Fig. 3.2), an exercise is proposed that requires careful reading; the point P(P′, P_{xy}') does not always and only occur in the first quadrant; it is chosen with a negative coordinate according to x, y, z, creating an orientation problem.

Now, let us conduct for P a line a(a′, a_{zx}) parallel to y; to locate the track on zx, point 1 must be located there; on the T_{azx} is the a_{zx}; moreover, let the projecting line b(b′, b_{xy}') incident x at the point Q be conducted through P; determine P-Q.

The distance P_{xy}'-Q_x is the hypotenuse P_{xy}'-2 of P_{xy}'-1 and 1-Q_x; the hypotenuse P′-3 of P_{xy}'-P_{xy}'-3 = P_{xy}'-2 is the distance P-Q; the angle formed by b with xy is given by P_{xy}'-3-P′.

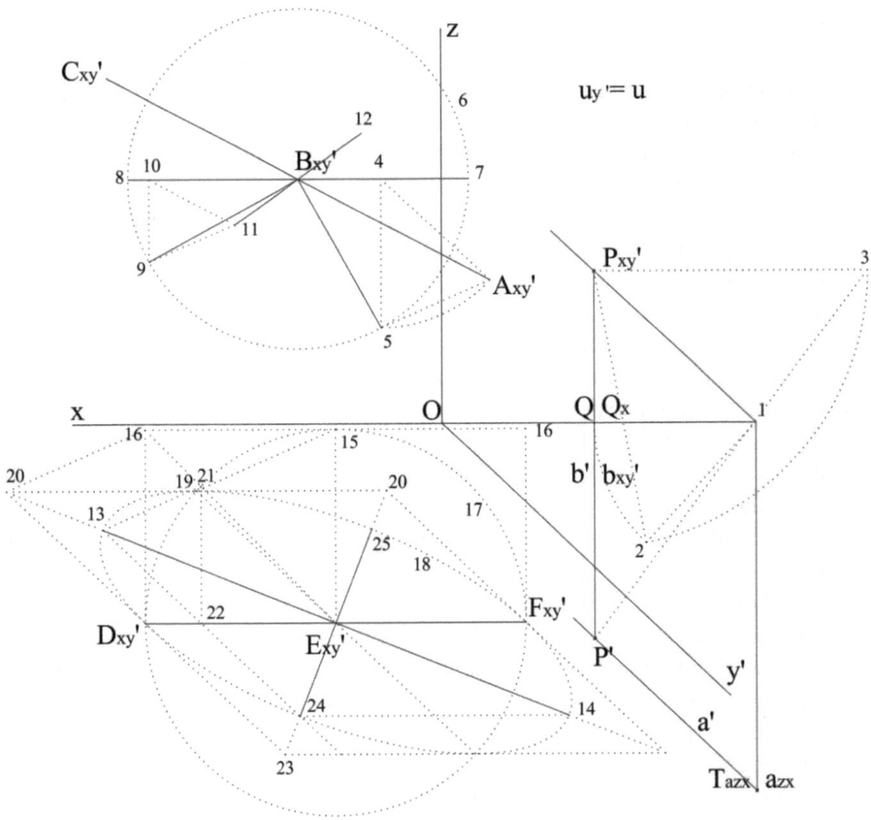

Fig. 3.2 Distances and slope angles: Measuring real distances and inclinations between spatial elements

A segment given on the picture $\pi = xz$, however, direct, retains its size; if it is given on xy, its size varies with the direction; if it is parallel to x or y′, its projection has the real size, at the graphic scale adopted on the picture.

If it rotates, approaching the direction of the long diagonal of the rhombus, the projection of the square with sides parallel to the two axes, its size increases and is maximum if parallel to it; it is perceptible that in this direction there is a maximum stretching.

Conversely, if it rotates, approaching the direction of the short diagonal of the aforementioned rhombus, the size decreases and becomes minimal when it assumes this direction; there is maximum contraction.

It should be remembered that the diagonals of the rhombus are orthogonal, both in reality and in projection, and represent the directions of the axes of the ellipse, a projection of the circumference inscribed in the square, identified by the aforementioned rhombus.

Then, let the A_{xy}'-B_{xy}'-C_{xy}' projection be given of a segment of xy with a generic direction; it must be reiterated that reading and reflecting on symbols is essential, to avoid fallacious readings. The segment is on XY and not on XZ, as it might seem, out of recklessness. Always remember that appearing is as different as being.

Obtain the true dimension B_{xy}'-5 of B_{xy}'-A_{xy}', through A_{xy}'-4 parallel to y′ and 4-5 orthogonal to x; in the circumference of radius B_{xy}'-5, 5 corresponds to A_{xy}' in the orthogonal affine homology of axis 7-4-B_{xy}'-8.

If we want to identify the size of the given segment in a different direction, for example, according to the direction conjugated to A_{xy}'-B_{xy}'-C_{xy}', consider B_{xy}'-9 orthogonal to B_{xy}'-5 and, through the 9-10 and 10-11 parallel to 5-4 and 4-A_{xy}', we obtain, with 9-11 parallel to 5-A_{xy}', the direction of the improper center of homology, the extreme 11 of 11-B_{xy}'-12, projection of the given segment, rotated in the desired direction.

The given segment is longer than the real, the conjugate is shorter; the two real segments are orthogonal to each other, the projections are not, unless the direction of the first, by chance, is that of the diagonal of the rhombus; in this case the given segment is the major axis, i.e. it has the maximum length and the conjugate is the minor axis.

If I notice a horizontal segment D_{xy}'-E_{xy}'-F_{xy}', if we want to construct the projection 13-E_{xy}'-14 that appears as long as possible, that is, that it is directed according to the long diagonal of the rhombus projection of the square on the side equal to 13-E_{xy}'-14 and parallel to x, or that it turns out to be the major axis of the ellipse homologous to the circumference inscribed in the aforementioned square.

Consider the quarter of the square, D_{xy}'-E_{xy}'-15-16 and the circumference 17 of center E_{xy}' and diameter D_{xy}'-E_{xy}'-F_{xy}', inscribed in the entire square; its projection is ellipse 18 linked to the aforementioned circumference by oblique affine homology, having the axis in D_{xy}'-E_{xy}'-F_{xy}' and the improper center in the direction 15-19, being corresponding and therefore homologous as averages of the side of the square and of the rhombus.

The homologue 13 of 21, intersection between E_{xy}'-16 and 17, is obtained through the sequence 21-22, 22-13, 21-13; the 13 and 14, symmetrical with the 13 with respect to the E_{xy}' center, identify the major axis; the 13-22 intercepts the 24 on the other diagonal, which with the symmetrical 25 defines the minor axis 24-E_{xy}'-25; Once the axes are known, we obtain ellipse 18.

Once the given segment has been established in a new direction in projection, the diameter of that direction is considered, the size of which is that assumed in projection by the given segment; the direction on the homologous circumference can be considered, and the homologous diameter can be obtained from it, and continue as already mentioned.

The ellipse 18 can be traced with a good approximation, by performing the aforementioned construction; from it can be drawn light of apparent lengthening or shortening of visual perception, acceptably controlled by rational perception; For different lengths of the assigned base segment, we operate by proportions, i.e. according to the rules of homothety, both for 18 and for the diameter of the required direction.

With reference (Fig. 3.3) to the projection O-x′, y, z, with $u_x' = u$, for a point P out of range, identified by the intersection of the line a(a′, a_{xy}', T_{axy}', T_{azx}') with the line b(b′, b_{xy}', T_{bxy}', T_{bzx}'), we want to conduct the line d parallel to the line c(c′, c_{xy}', T_{czx}').

Two incident lines are coplanar and the trace of their plane on xy is the T_{axy}'-T_{bxy}' and the one on zx is the T_{azx}'-T_{bzx}'; The two tracks meet on X′, off-screen; the two lines cannot both be given; given the a, of the trace Ta_{zx}' defined by point 1, of the

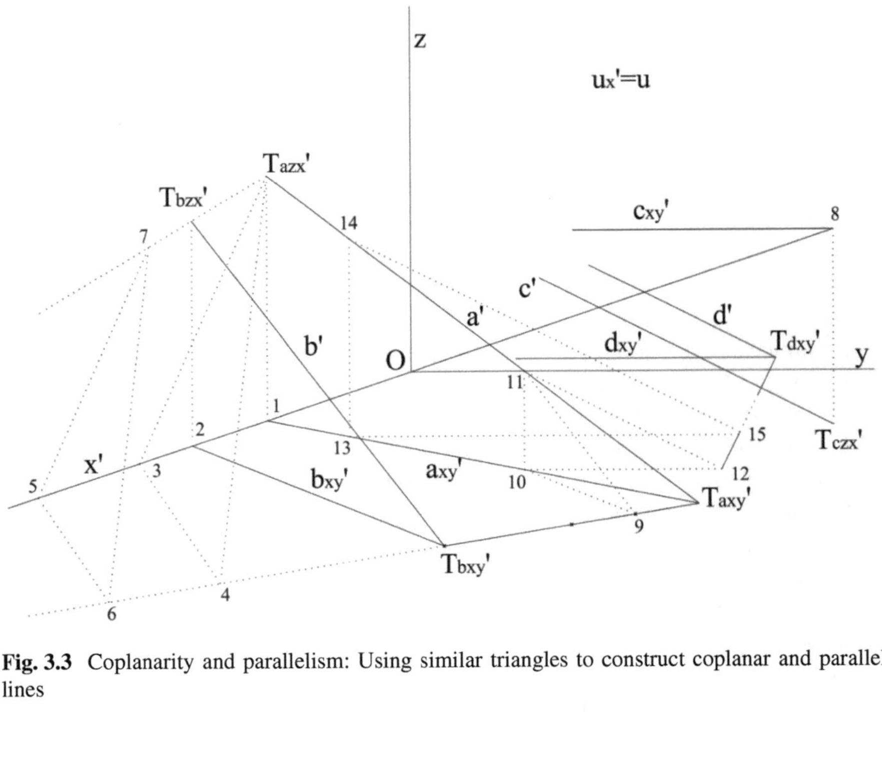

Fig. 3.3 Coplanarity and parallelism: Using similar triangles to construct coplanar and parallel lines

b are chosen the T_{bxy}', the b_{xy}', which on x′ identifies the 2, and the b′, but not the T_{bzx}'.

To ensure coplanarity and therefore the point P of intersection out of range, x′ and T_{axy}'-T_{bxy}' must be considered; the T_{azx}'-3-4 triangle is built and flanked by the homothetical 5-6-7; finding the 7, you get the T_{azx}'-7 and on it the T_{bzx}'.

Once the coplanarity has been achieved, it is necessary to hook P with the line d, parallel to c; divide the T_{axy}'-T_{bxy}' section into equal parts, for example into four parts, identifying 9; for the 9th the 9-10 and 9-11 are conducted parallel to b_{xy}' and b′; for the 10th and 11th the 10-12 and 11-12 parallel bars are drawn at c_{xy}' and c′.

Doubling T_{axy}'-10 in T_{axy}'-13 gives the second triangle 13-14-15, and then of the fourth triangle, which brings the 14 to P′, we get the R_{dxy}', the d_{xy}', and the d′; reflect on the essential use of similitude.

The importance and the essential need to dust off the old and distant knowledge of Elementary Geometry from oblivion are highlighted, adapting the role and treatment of each of its concepts at the level of the teacher, for categorical commitment.

It should also be noted that the choice of a repertoire of proposed resolutions, which, invariably, forces an inevitable series of profound observations of the data.

Interpretations and reflections, reasoning and preventive convictions at the beginning of the choices of operational procedures, are an expression and sign of responsible maturity, rewarded, in the end, by the conscious acquisition of skills, dynamism, safety, merit, reliability, trust, largely gratifying.

Consider the representation of the planes (Fig. 3.4); the projection of the reference O, x, y, z′ is given, with $u_z' = u$; On a) we read, in addition to the coordinated planes xy, yz′, z′x, several planes of generic and particular position.

The plane α ($t_{\alpha zx}'$, $T_{\alpha z}'$, $t_{\alpha yz}'$) is parallel to xy; the distance of α from xy is the dimension of any of its elements; the trace taxy is infinite; the dimension can be positive or negative; α is parallel to x, to y, and to any line of xy; α is orthogonal to zx and yz.

The plane β ($t_{\beta zx}'$, $T_{\beta x}$, $t_{\beta xy}$) is parallel to yz, y, z′, and orthogonal to xy and zx; the trace $t_{\beta yz}'$ is improper; its distance from yz can be positive or negative.

The plane γ ($t_{\gamma yz}'$, $T_{\gamma y}$, $t_{\gamma xy}$) is parallel to xz, x, z′, and orthogonal to xy and yz; the trace $t_{\gamma zx}'$ is improper; its distance from xz can be positive or negative.

The plane δ ($t_{\delta zx}'$, $T_{\delta x}$, $t_{\delta xy}$) is parallel to z′, and orthogonal to xy; the trace $t_{\delta yz}'$ is out of range and parallel to z′; the O-$T_{\delta x}$ can be positive or negative; the angle between δ and xz is read at xy = p in x-$T_{\delta x}$-$t_{\delta xy}$, independent of the scale of representation u.

The plane ε ($t_{\varepsilon zx}'$, $T_{\varepsilon x}$, $t_{\varepsilon xy}$, $T_{\varepsilon y}$, $t_{\varepsilon yz}'$) is parallel to z′, and orthogonal to xy; the trace $T_{\varepsilon z}'$ is improper; the O-$t_{\varepsilon xy}$ can be positive or negative; the angle between ε and xz and between ε and yz are read in real on xy = π in x-$T_{\varepsilon x}$-$t_{\varepsilon xy}$, and in y-$T_{\varepsilon y}$-$t_{\varepsilon xy}$ independent of the scale of representation u; a figure of ε in projection retains shape and size, since $t_{\varepsilon xy}$ is parallel to π and $u_z' = u$.

A figure of α is preserved in shape and size; a figure of β, γ, δ, in projection is obtained by homology from the one in true form and size placed on the given plane, overturned on xy = p by rotation around $t_{\beta xy}$, $t_{\gamma xy}$, $t_{\delta xy}$; the second trace is reversed orthogonally to the first, which is the axis of rotation.

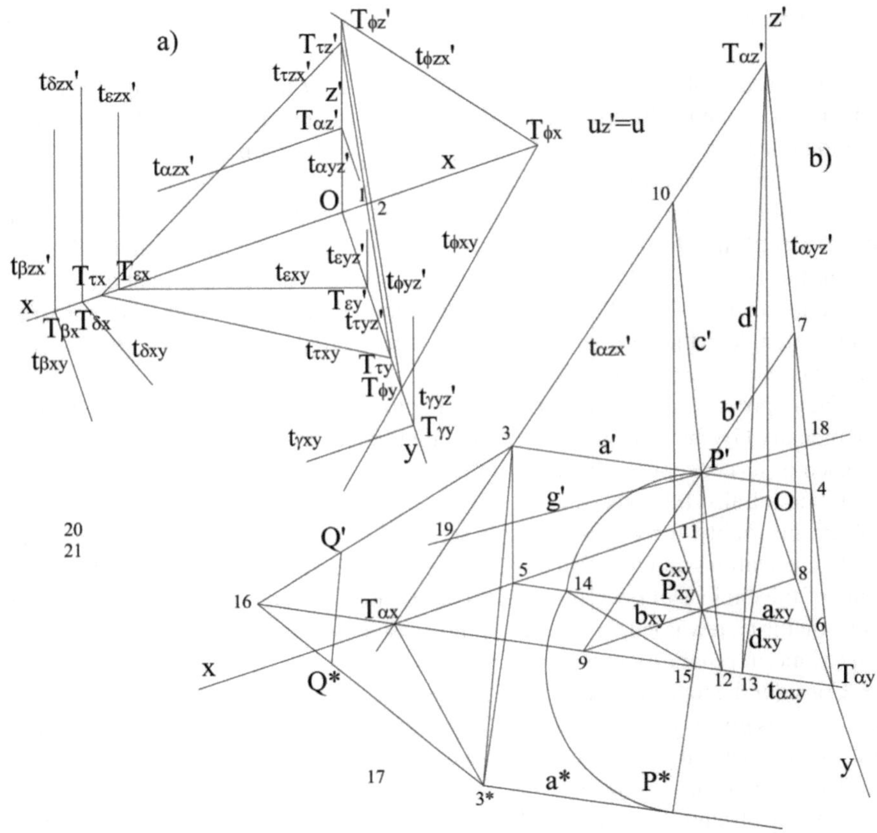

Fig. 3.4 Planes and their projections: **a** The plane α is parallel to xy, β is parallel to yz, γ is parallel to xz, δ is parallel to z′, and orthogonal to xy, ε is parallel to z′, and orthogonal to xy. If the traces of two generic planes $\tau(t_{\tau xy}, T_{\tau x}, t_{\tau zx}', T_{\tau z}', t_{\tau yz}', T_{\tau y})$ and $\varphi(t_{\varphi xy}, T_{\varphi x}, t_{\varphi yz}', T_{\varphi y}, t_{\varphi zx}', T_{\varphi z})$ are moved along the axes, the planes change position when viewed from above (projecting downward), and so on; **b** on the plane $\alpha(t_{\alpha xy}, T_{\alpha x}, t_{\alpha zx}', T_{\alpha z}', t_{\alpha yz}', T_{\alpha y})$, consider the point $P(P', P_{xy})$ and note that among the infinite lines of α passing through P, some can be considered special due to their direction

Read carefully the two generic planes $\tau(t_{\tau xy}, T_{\tau x}, t_{\tau zx}', T_{\tau z}', t_{\tau yz}', T_{\tau y})$ and $\phi(t_{\phi xy}, T_{\phi x}, t_{\phi yz}', T_{\phi y}, t_{\phi zx}', T_{\phi z})$; bring O-$T_{\tau z}'$ orthogonal to the sheet in O and imagine $T_{\tau z}$ joined with $T_{\tau x}$ and $T_{\tau y}$; the plane τ is seen from above.

Move $T_{\tau x}$ to position 1 on x; the plane becomes projecting; continuing the movement of $T_{\tau x}$, the other side of the plane τ is seen; think $T_{\tau x}$ to infinity; reflect on what happens.

Passing from one side of a finite point to the other, one can say "straddling the point"; for the point to infinity, it loses meaning, because it cannot have a location; Infinity cannot be understood, it remains a mystery, beyond reason.

As a consequence, for the same position, one sees one or the other face of the plane τ indifferently; on the other hand, it is clear that one understands the difference of seeing the plane from above or from below.

This is also true for the generic plane ϕ, point 2 indicates for $T_{\phi x}$ the projecting position; we go further and change the visible side; there is something that recalls the singular property of the Moebius strip, a single-sided surface.

Reading (b), on the plane $\alpha(t_{\alpha xy}, T_{\alpha x}, t_{\alpha zx}', T_{\alpha z}', t_{\alpha yz}', T_{\alpha y})$ consider the point P(P′, P_{xy}); the point belongs to the plane if it belongs to a line of the plane; the planes containing P are infinite; the lines containing P are infinite; if P belongs to a plane, the lines on α containing P are infinite; geometrical reason cannot discuss the apparent differences of the infinities mentioned.

A line belongs to a plane if its tracks belong to the homonymous tracks of the plane; A point belongs to a line of the plane if the projections of the point are on the projections of the same name.

Among the infinite lines of α, containing P, some can be said to be particular because of their direction; a (a′, a_{xy}) is parallel to xy, to taxy, to a_{xy}; a′ contains P′, a_{xy} contains P_{xy}; with the same reading criterion, b is parallel to xz, with 9-8 = b_{xy} parallel to x; c′ is parallel to y-z, with 11-12 = c_{xy} parallel to y.

For O, consider O-13 = d_{xy}, orthogonal to the $T_{\alpha xy}$, and $T_{\alpha z}'$-13 = d′; on α d is of maximum slope on π; if we want to flip α on π; rotating α around the line common to α and π, i.e. around the $T_{\alpha xy}$, the point P describes an arc of circumference on the plane orthogonal to the $T_{\alpha xy}$, whose track on xy is P_{xy}-15; the radius of the arc P-15 is the hypotenuse 15-P which is obtained in 14-15, after carrying the P′-P_{xy} in dimension P_{xy}-14 orthogonal to P_{xy}-15.

The point P rotating lies on π, which in this case is xy, in P*, with 15-P* = 15-14 = 15-P; reflect on the distinction of the centers of the two marked arcs; P* is homologous to P′ in the oblique affine flipping homology defined by the $t_{\alpha xy}$ axis, the pair P′, P*, and the improper center P′-P*.

Corresponding lines contain corresponding points, aligned with the center P′-P* and meet on the axis; the line a′ for P′ is parallel to the axis taxy, corresponds to the line a* for P* and parallel to taxy; the 3 of a′ on the 5 of a_{xy}, corresponds 3* on the a*; the $T_{\alpha x}$-3 is homologous with $T_{\alpha x}$-3*; the 3-16 corresponds to the 16-3*; the point Q′ on 3-16 of α is homologous Q* on 16-3* of α*; from the generic line of α g′ = 17-P′-18, through 17-17*, we obtain g* = 17*-P*; reflect, again, on the homologous lines Q′-R′ and Q*-R*, whose joined point is 19.

Once the vertices of a figure σ* are assigned on α*, if the relative projection σ' is desired, a P(P′, P_{xy}) of α is chosen, the homologous pair P′, P* is defined and by homology the projection of the vertices of the corresponding s′ is obtained, followed by the projection of the vertices of σ_{xy}.; observe the usefulness of leaning freely on any pair of points and homologous lines to obtain the homologous element of the one under consideration.

If σ' is known and the true shape and magnitude σ* is desired, the reversal of an appropriate vertex of σ' is obtained, i.e., any P(P′, P_{xy}) of α is considered, the pair P′-P*, which defines homology, is obtained, and the σ* is obtained as above.

The solution is not a standard, but a continuous test of intelligent choices, linked to experience, guided by the sense of challenge and confrontation in the search for gradually more and more complicated hypotheses; for example, if $T_{\alpha x}$ on x is out of bounds, what happens? It is proposed, and the winning process is addressed.

A few more examples (Fig. 3.5) of planes with less frequent positions: in (a), a point P and a straight line are given of the plane a; generically, when the plane is defined by a line and a point, the given line is considered for the point, or more simply the parallel; by joining the homonymous traces of the two lines, the homonymous traces of the plane are obtained.

If the line is an axis, for example, x, for P, we lead the parallel to x and we obtain P_{yz}'; $t_{\alpha xy}$ is superimposed on $t_{\alpha zx}$ and x, while O-P_{yz}' is $t_{\alpha yz}'$; it can be observed that without a point P, the plane for the axis is not defined, although two traces of the plane are given.

Also in (a), there is the plane β, to which O and the line r′(r′, T_{rxy}, T_{ryz}') belong; r′ appears parallel to z′, but r and z are not parallel at all; it is necessary to read very carefully and reflected; think of conducting the parallel for O; the three traces of this line are in O; in particular the two homonymous to those of r are in O; the traces $t_{\beta xy}$ and $t_{\beta yz}'$ of plan β are obtained in O-T_{rxy} and O-T_{ryz}'.

The third track $t_{\beta zx}'$ of plan β contains O and is parallel to 1-2, which joins the projection on z′ of T_{ryz}' and the projection on x of T_{rxy}; 1-2 is r_{zx}'; it is reiterated that the correct interpretation of points, lines and planes, considered by many to be a trivially easy operation, can present unexpected pitfalls.

In (b) the data to construct the plane α are the trace $t_{\alpha yz}'$, of which only the $T_{\alpha z}'$ is in the field, and the projections P′ and P_{xy} of P; for P′ we conduct the r′ parallel to the $t_{\alpha yz}'$ and for P_{xy} the parallel to y; once the traces 1 and 3 have been identified, we construct the $t_{\alpha zx}' = T_{\alpha z}'$-3-4 and the $t_{\alpha xy} =$ 4-1.

In (c) the data for determining the plane γ are the three points P(P′, P_{xy}), Q(Q′, Q_{xy}), R(R′, R_{xy}); for two of them the line is conducted and for the third the parallel; the relative traces derive the traces of the plane; the choice of the first two points must be prudent and not give rise to avoidable complications.

The reading of both the data and the elements of the solution process must always be very careful; the P′-Q′ is randomly parallel to y, but the P-Q is not parallel to y at all; it is detected by P_{xy}-Q_{xy}.

The sequence of operations is: P′-Q′, P_{xy}-Q_{xy}, 1, 2 on x, 3, R_{xy}-4 parallel to P_{xy}-Q_{xy}, R′-5 parallel to P′-Q′, the 5-3 $= t_{\gamma zx}'$, $T_{\gamma z}'$, $T_{\gamma x}$, $T_{\gamma x}$-1,$T_{\gamma y}$, $T_{\gamma y}$-$T_{\gamma z}'$.

In (d) the intersection $i_{\alpha\beta}'$ between two planes α($t_{\alpha xy}$, $T_{\alpha x}$, $t_{\alpha zx}'$) and β($t_{\beta xy}$, $T_{\beta x}$, $t_{\beta zx}'$) is proposed; the difficulty lies in the reading of $i_{\alpha\beta}'$, disoriented by its parallelism to z′; $i_{\alpha\beta}$ is not parallel to z at all; this is made clear by the projection of its projection 1-2 $= i_{zx}'$.

In (e) another case of reading is presented, which impacts perplexity; the intersection igd concerns the plane γ($t_{\gamma xy}$, $T_{\gamma y}$ out of range on y, $t_{\gamma yz}'$) and the plane δ($t_{\delta zx}'$, $T_{\delta z}'$ out of view on z′, $t_{\delta yz}'$); yes of indeed traces of planes incident on y′ and z′ respectively.

By means of homothetic triangles, not shown in the figure, this quality is ascertained; there are also T_{iyz}' and $T_{\delta y}$; by means of the dotted homothetic triangles in

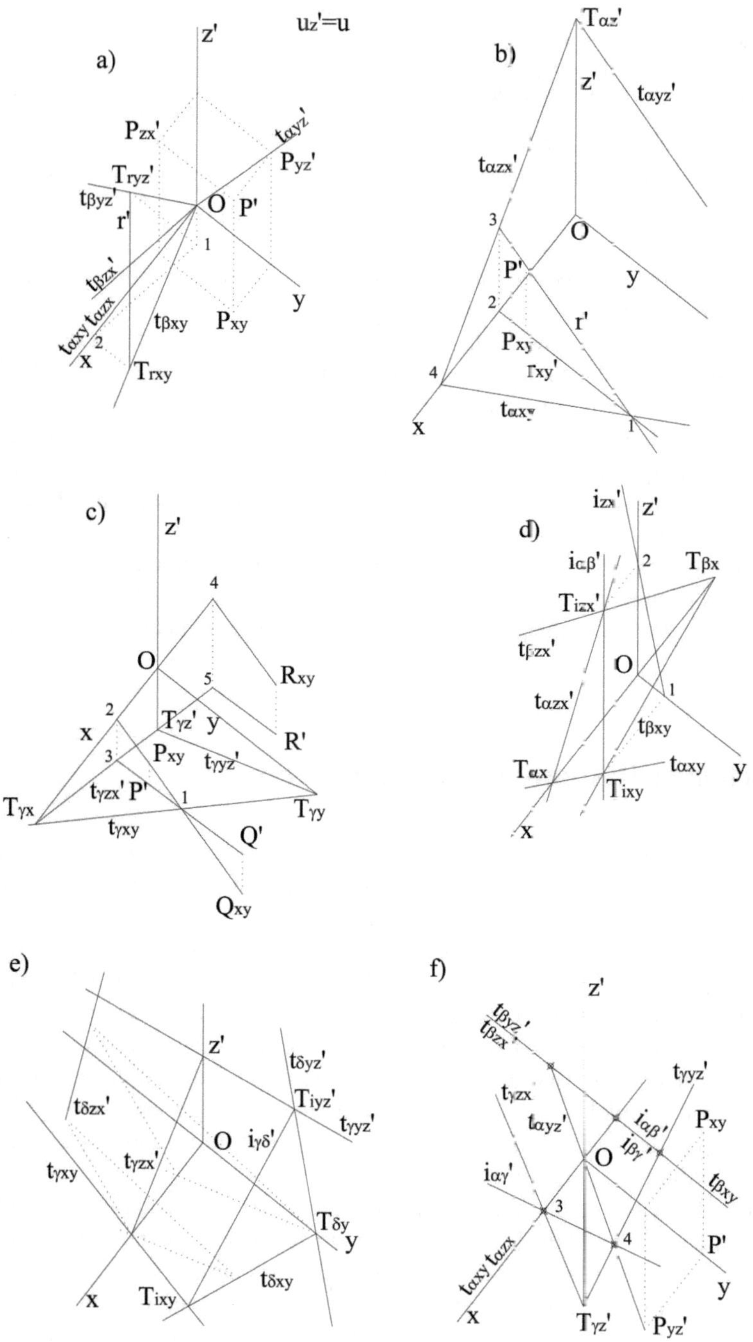
a)
b)
c)
d)
e)
f)
uz'=u
z'
O
x
y
Pzx'
Pyz'
P'
Pxy
Tryz'
Trxy
r'
Tαz'
tαyz'
tαzx'
Tγx
Tγy
Tγz'
tγxy
tγzx'
tγyz'
R'
Rxy
Q'
Qxy
Tβx
Tαx
Tixy
Tizx'
tβzx'
tαzx'
tβxy
tαxy
tδzx'
tγxy
tγzx'
tδyz'
Tiyz'
tγyz'
iγδ'
Tδy
tδxy
Tγz'
Pyz'

◂**Fig. 3.5** Constructing planes from points and lines: Defining planes via point-line combinations with homological methods. **a** plane β, to which O and the line r′(r′, T_{rxy}, T_{ryz}') belong; r′ appears parallel to z′, but r and z are not parallel at all; **b** given α for P′ we conduct the r′ parallel to the $t_{\alpha yz}'$ and for P_{xy} the parallel to y; **c** given three points P(P′, P_{xy}), Q(Q′, Q_{xy}), R(R′, R_{xy}) we can conduct the plane γ; given two planes α($t_{\alpha xy}$, $T_{\alpha x}$, $t_{\alpha zx}'$) and β($t_{\beta xy}$, $T_{\beta x}$, $t_{\beta zx}'$) we can find the intersection; **e** use of homothetic triangles; **f** given three planes α($t_{\alpha xy}$ = x = $t_{\alpha zy}$, P′, P_{xy}'), the plane β($t_{\beta xy}$ = $t_{\beta yz}'$ = $t_{\beta zx}'$ parallel to y′, projecting), and the plane γ($t_{\gamma yz}'$, $T_{\gamma z}'$, $t_{\gamma zx}'$) we can find the point of intersection

the figure, the $T_{\delta xy}$ is obtained, incidentally, the $T_{\delta zx}'$ in $T_{\delta x}$ on x off-screen; finally, with T_{ixy}, we obtain $i_{\gamma\delta}' = T_{iyz}'\text{-}T_{ixy}$.

In (f) the common point of three planes is sought, which do not belong to an improper bundle of planes; the data predict the plane α($t_{\alpha xy}$ = x = $t_{\alpha zy}$, P′, P_{xy}'), the plane β($t_{\beta xy}$ = $t_{\beta yz}'$ = $t_{\beta zx}'$ parallel to y′, projecting), and the plane γ($t_{\gamma yz}'$, $T_{\gamma z}'$, $t_{\gamma zx}'$).

The intersections $i_{\alpha\beta}'$ and $i_{\beta\gamma}'$ overlap on the traces of projecting β, and the intersection $i_{\alpha\gamma}'$, defined by points 3, 4, is the point, off-screen, common to the three planes; it is necessary to reflect and not to underestimate, as happens, the exercise with an apparently simple and easy text.

Often, the generic statement of elementary geometric problems can lead to rash conclusions of unequal comparisons with other, more feared ones; it is a prejudice that stains the growth of the prejudiced; wanting to understand and reflect can avoid all this.

Projection or Restitution of a Plane Figure

In Fig. 3.6a, the solution process of the projection of a given σ on xz is illustrated; xz is reversed on xy = p; z rotates to O with respect to x, becomes z*, and overlaps y.

Once the true shape and magnitude of the σ is placed in σ*, the A*-A_x parallel to z*, being $u_z' = u$, projects in true length in A_x-A′ parallel to z′; A* and A′ are related by oblique affine homology, defined by the x-rotation axis and the center A*-A′.

Observing the σ′ obtained from σ*, in the transformation, we detect some qualities that persist, others that vary; a side parallel to x retains the direction and dimension.

For a side parallel to z*, the size and direction change are conserved while remaining parallel to z′; the angle between a side parallel to x and a side parallel to the center of homology A*-A′ retains its value.

In any other case, neither the direction nor the dimension is conserved between a linear element of σ* and the homologue in σ′; in any other case, between an angular element of σ* and the homologue in σ′, the measure is not conserved.

Two parallel sides remain parallel; there is a direction that transforms with the maximum shortening and another with the maximum elongation; A regular figure does not remain such.

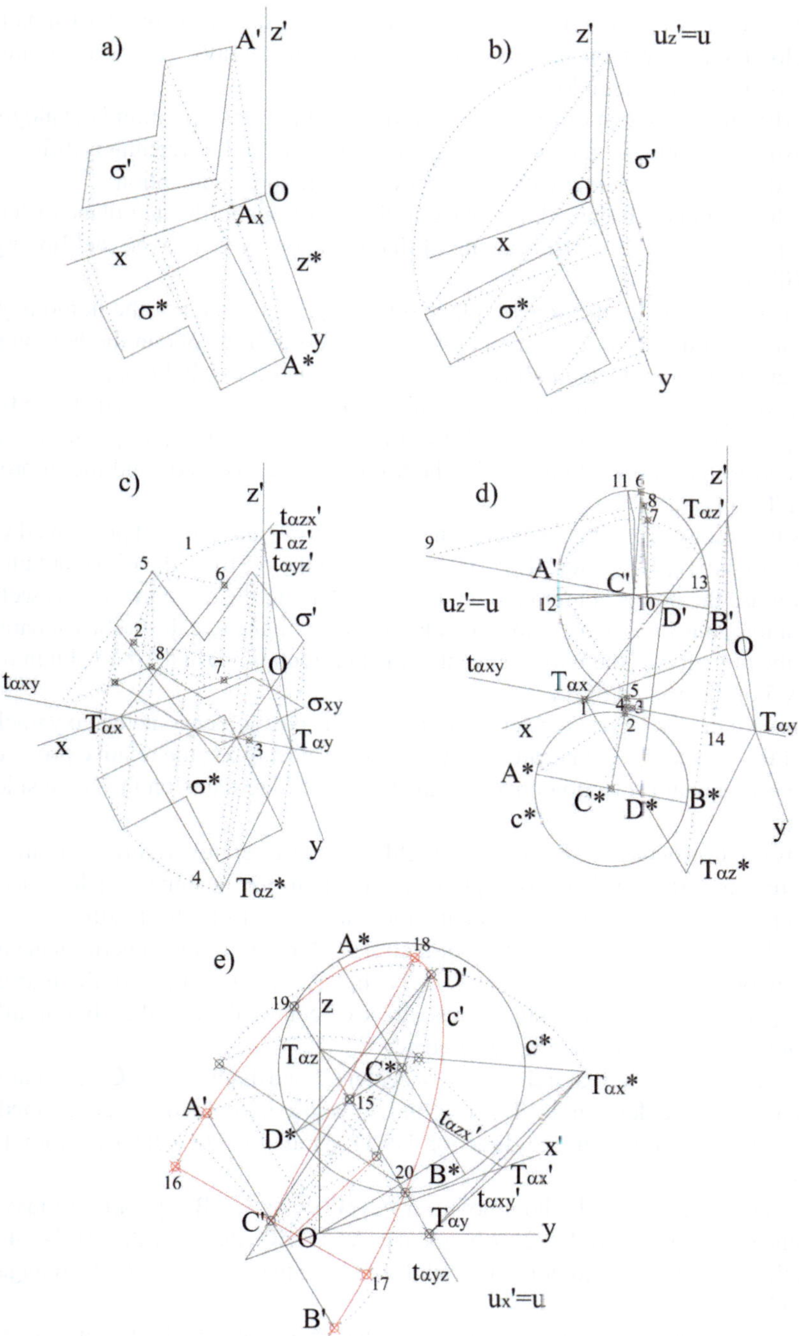

Fig. 3.6 Projection of planar figures: Restitution of figures using axis and homology reversal. **a** projection of a given σ on xz; **b** σ belongs to yz; **c** σ belongs to the generic plane α ($t_{\alpha xy}$, $t_{\alpha zx}'$); **d** projection of a circle belonging to the generic plane α; **e** representation on the reference x′, y, z, O, $u_x' = u$, $t_{\alpha xy}'$, $t_{\alpha yz}$, $t_{\alpha zx}'$, $T_{\alpha x}'$, $T_{\alpha y}$, $T_{\alpha z}$

The joining of two vertices and the homologous joining of the corresponding homologues intersect on the axis; at a center point of rays, the center point of homologous rays is homologous.

In (b), the σ belongs to yz; z is reversed on x; the transformation is analogous; the visual perception of the image obtained is different, and its reading is differently demanding, commensurate with the experience of rational perception.

In the examples discussed, the plane in figure σ is a coordinated plane different from the picture $\pi = xy$; the drawing of that plane on xy is an axis; the flipping is simplified.

In (c) σ belongs to the generic plane α ($t_{\alpha xy}$, $t_{\alpha zx}'$); the axis of the homology of reversal of α on xy is $t_{\alpha xy}$; of the point $T_{\alpha z}(T_{\alpha z}', O)$ of α we obtain the homologue $T_{\alpha z}$*; arc 1 of center O and radius O-$T_{\alpha z}'$, we obtain O-2 parallel to $t_{\alpha xy}$.

The orthogonal for O to the $t_{\alpha xy}$ identifies 3 on $t_{\alpha xy}$; the 2-3 is $T_{\alpha z}$'s distance from the $t_{\alpha xy}$; arc 4 of center 3 and radius 2-3 = $T_{\alpha z}$-3 gets $T_{\alpha z}$* on O-3; the points $T_{\alpha z}'$ and $T_{\alpha z}$* correspond in the oblique affine homology of the taxy axis and the improper center $T_{\alpha z}'$-$T_{\alpha z}$*.

By homology, from the σ reported in σ* in true form and size and positioned with respect to the $T_{\alpha x}$*-$T_{\alpha z}$* and $T_{\alpha y}$*-$T_{\alpha z}$* we obtain the σ′; from the σ′ we obtain the σ_{xy}; we use the points common to the $t_{\alpha xy}$, to the σ*, to the σ′, to the σ_{xy}, as well as the homologous lines of homologous sides, incident on the $t_{\alpha xi}$, the verticals, parallel to z′, the direction of the improper center, the horizontals 5-6 and 7-8 to obtain from vertex 5 of σ′ vertex 8 of σ_{xy}.

Observe the passage of the vertex of σ* from the position beyond the taxy track to that of the vertex of σ′, at negative altitude with respect to xy; the relative projection of the orthogonal projection on xy of this vertex of σ′ remains on the same side of the $t_{\alpha xi}$.

Any rotation of the σ on its plane can highlight or exasperate certain appearances; it is interesting to deepen these conceptual investigations for the impact on the practical perceptual results, both in the visual and rational sense, useful for reading.

In (d) we propose the projection of a circle belonging to the generic plane α of the previous exercise; given x, y, z′, O, $u_z' = u$, $t_{\alpha xy}$, $T_{\alpha x}$, $T_{\alpha y}$, $T_{\alpha z}'$, we flip α around the $t_{\alpha xy}$ on $xy = \pi$, identifying, as has already been explained, the oblique affine homology, of taxy axis and improper center $T_{\alpha z}'$-$T_{\alpha z}$*.

Position on α* the circumference c*, of center C* and diameter A*-C*-B*, parallel to the $t_{\alpha xy}$ and incident on the $T_{\alpha x}$-$T_{\alpha z}$* in D*, whose homologue D′ is obtained on the $T_{\alpha x}$-$T_{\alpha z}'$, according to the direction D*-D′ parallel to the center of homology $T_{\alpha x}$-$T_{\alpha z}$*.

For D′, we consider the homologue A′-C′-B′ = A*-C*-B*, parallel to taxy; in sequence, draw the radius C*-2, orthogonal to A*-C*-B*, the A*-2-3, the C*-2-4, the 4-C′, the 3-A′, the 5-C′ homologous radius of C*-2, the diameter 5-C′-6 conjugated to A′-C′-B′.

Once the pair of conjugate diameters has been obtained, the ellipse c′ is constructed; among the different ways to obtain it, it is worth mentioning the orthogonal affine homology that links the circumference of center C′ and diameter A′-C′-B′ to the ellipse c′.

Having computers, the construction is carried out with the command that provides for the availability of the axes of the c′; the axes can be obtained by means of the sequence: circumference of center C′ and diameter A′-B′, C′-7 orthogonal to A′-B′, 6-7, its midpoint 8, 8-9 orthogonal to 6-7, arc of center 9 and radius 9-6 = 9-7 by obtaining 10, homologues 6-10 and 10-7.

By translating the group 6-10-7 to C′, we determine the end 11 of the semi-major axis parallel to 10-6; by translating the group orthogonal to the previous one, the end 12 of the minor axis 12-C′-13 is identified; with the Autocad command, you get the ellipse c′.

In a backward procedure, of restitution, it can be verified whether, given any ellipse c′, it represents a circumference of xy; the diameter A′-C′-B′ parallel to the $t_{\alpha xy}$ must be conjugated to the diameter 5-C′-6, parallel to the projection T_{az}'-14 of the line of maximum slope of α with respect to xy = π, O-14 being orthogonal to $t_{\alpha xy}$.

This is not enough; it is necessary to derive the true length of 5-C′-6, which must be equal to A′-C′-B′; through the "visual" analysis, the reasons for certain exclusions are more easily derived. An affirmative answer requires all of the aforementioned checks.

In (e) there is a representation on the reference x′, y, z, O, $u_z' = u$, $t_{\alpha xy}'$, $t_{\alpha yz}$, $t_{\alpha zx}'$, $T_{\alpha x}'$, $T_{\alpha y}$, $T_{\alpha z}$; the plane α flips on yz = π by rotation around the t_{ayz}; the point T_{ax}' of x′, in space is on the orthogonal in O to yz at the distance C-15 = O-Tax′.

The 15-16, the inverted distance of Tax from the $t_{\alpha yz}$, rotates and obtains $T_{\alpha x}$* on the O-16 orthogonal to the $t_{\alpha yz}$; the triangle $T_{\alpha x}$*-$T_{\alpha y}$, $T_{\alpha z}$ identifies α*; the homology of reversal is defined by the $t_{\alpha yz}$ axis and the improper center $T_{\alpha x}$*-$T_{\alpha x}'$ and by the triangle $T_{\alpha x}$*-$T_{\alpha y}$-$T_{\alpha x}'$-$T_{\alpha x}$*.

The c* of center C* and diameter A*-C*-B* parallel to $t_{\alpha yz}$ is positioned; the homothetic triangle to the aforesaid and having vertex in C*, rests on the $t_{\alpha yz}$ according to the direction $T_{\alpha x}$*-$T_{\alpha y}$; from this point draw the parallel to the $T_{\alpha y}$-$T_{\alpha x}'$, until the line for C* parallel to the $T_{\alpha x}$*-$T_{\alpha x}'$ intersects in C′.

For C′ write A*-C*-B* in A′-C′-B′, diameter of c′; to the radius D*-C* orthogonal to A*-C*-B*, incident the $t_{\alpha yz}$ in 15 corresponds to the conjugate semi-diameter C′-15-D′, D′ being the intersection between C′-15 and D*-D′ parallel to the $T_{\alpha x}$*-$T_{\alpha x}'$; The availability of the pair of conjugate diameters defines the ellipse c′.

The axes can be derived by means of the construction already illustrated; the minor axis is 16-C′-17 and the semi-major axis is C′-18; note the inversion between c′ and c*; the extremes D′ to D* are placed by a band opposite to the axis of homology $t_{\alpha yz}$.

The radius of the overturning, which made it possible to obtain $T_{\alpha x}$* by clockwise rotation, can be carried out in the opposite direction; to this always follows the same c′; the aforementioned inversion does not occur.

D′ will be homologous to the opposite extreme of the current D* with respect to C*; the oblique affine homology related to the overturning with opposite rotation is different from the previous one; the axis is still $t_{\alpha yz}$, while the improper center will be the direction that unites $T_{\alpha x}'$ with the new position of $T_{\alpha z}$*.

Observe again that the c and the c* intersect on the $t_{\alpha yz}$ in 19 and 20; chosen for a point P* of the D*-C* of the c* any M*-P*-N* chord parallel to A*-C*-B*, the homologous M′-P′-N′ chord of the c′ is parallel to the $t_{\alpha yz}$ and of equal length.

The particular position adopted α leads to the result of an ellipse that is very elongated and largely out of range; differently imagined data produce results on which to reflect and be inspired by research requiring further investigation.

These are opportunities to experiment with speculative and operational strategies of great commitment and largely gratifying in the enrichment of skills, knowledge, and growth.

For example, it can be suggested to choose $\alpha = xy$; we can consider the circumference on this plane; the reversal of a is obtained by superimposing the spatial x-axis orthogonal to the frame yz on z by rotation around y.

Imagining that we give the inverted circumference tangent to both y and $x = z$ and considering the square circumscribed to the circumference and having the sides on the aforementioned axes, from this square derives, as an oblique axonometric projection, a rhombus, circumscribed to the ellipse projection of the circumference.

On reflection, the elementary observation follows that the diagonals of the square, orthogonal to each other, become diagonals of the rhombus, still orthogonal to each other; The two diagonals of the rhombus are the directions of the major and minor axes of the ellipse.

Another reflection, another conclusion: the circumference in projecting itself appears to deform as if it were subjected to a stretching force according to the major diagonal and compressing in the direction of the minor diagonal.

The ellipse is obtained by homology from the circumference inscribed in the square; the narrower the angle in O between x′ and y, the more the ellipse appears deformed, flattened; think of the diameter segment for its center.

This segment, in reality of invariable length, appears to vary from a minimum, having the direction of the minor diagonal of the circumscribed rhombus, to a maximum, having the direction of the major diagonal.

These directions and the associated values can also be identified by eye; in the two positions having the directions parallel to x′ and y, there is a real length.

Squareness and Distances

In Fig. 3.7a, we conduct a line n_α orthogonal to a plane $\alpha(t_{\alpha xy}', t_{\alpha yz}, t_{\alpha zx}', T_{\alpha x}', T_{\alpha z}$, improper $T_{\alpha y})$, generic and parallel to y, in the reference x′, y, z, with $u_x' =$ you and $yz = \pi$.

It overturns xz on yz; x′ overlaps as x* on y, rotating in O with respect to z; $T_{\alpha x}'$ is brought to $T_{\alpha x}$*; $t_{\alpha zx}'$ flips over to $t_{\alpha zx}$*; For a point 4 of the $t_{\alpha zx}$*, lead the n_α*, which intersects x* in 7.

Write the 4, of negative odds O-5 in 6 on $t_{\alpha zx}'$ and the 7 of x* in 8 on x′; 6-8 is n_α', generically orthogonal to α; given the point P(P′, P_{xy}'), we obtain P_{yz}, P_{zx}', through 1, 2, 3, and the orthogonal for P to α is $n_{\alpha P}'$, parallel for P′ to n_α'.

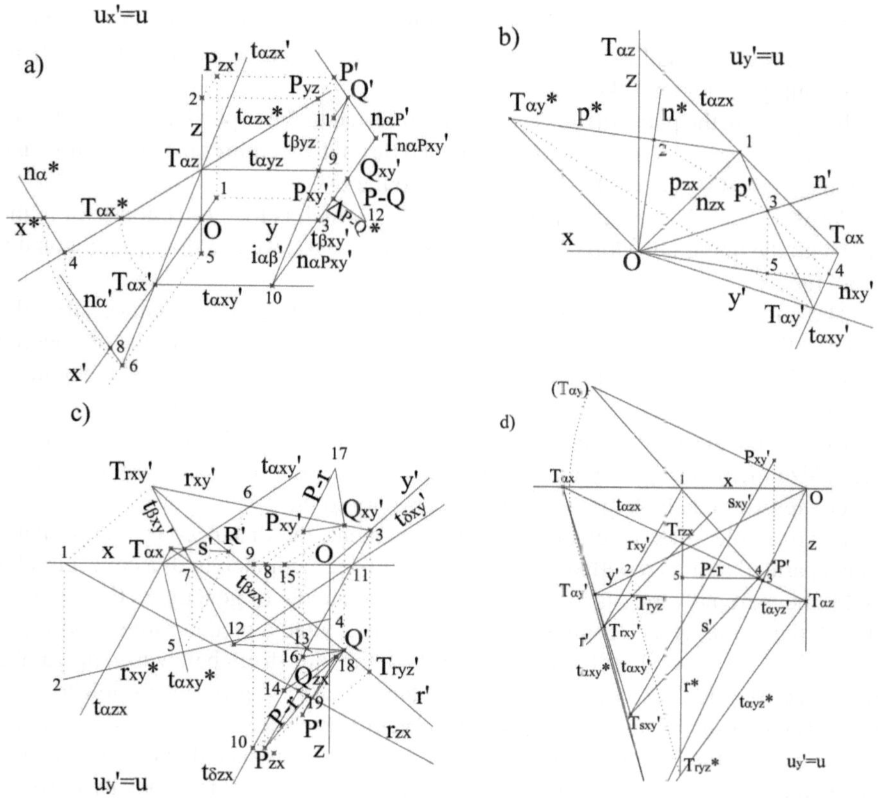

Fig. 3.7 Perpendiculars and spatial distances: **a** line n_α orthogonal to a plane α and point Q intersection between $n_{\alpha P}'$ and α and P-Q, the distance between the two points P and Q, as well as the distance of P from α; **b** Distance of P from α when $u_y' = u'$; **c** Distance of P from α when $xz = \pi$ and $u_y' = u$; **d** Distance of P from α when triplet x, y′, z, O and $u_y' = u$, in view from below

To determine Q (Q_α, $Q_{xy\alpha}$), the point of intersection between $n_{\alpha P}'$ and α, we consider the plane β ($t_{bxy\alpha}$, t_{byz}) for $n_{\alpha P}'$ parallel to xz; the intersection between $i_{\alpha\beta}$ is 9-10, which derives Q′ on $n_{\alpha P}'$; Q_{xy}' follows on $n_{\alpha Pxy}'$.

The $\Delta_{P\text{-}Q}^* = P_{xy}'\text{-}12 = 11\text{-}P'$, difference in altitude between P and Q, obtained by means of Q′-11 parallel to $n_{\alpha Pxy}'$, is referred orthogonally to P_{xy}'-Q_{xy}'; the Q_{xy}'-12 is P-Q, the distance between the two points P and Q, as well as the distance of P from α.

Note that the angle between α and xy is given by O-$T_{\alpha x}^*$-$T_{\alpha z}$ and between α and yz by O-$T_{\alpha z}$-$T_{\alpha x}^*$.

Any plane containing $n_{\alpha P}$ is orthogonal to α; the vertical for 3 identifies $T_{n\alpha Pyz}$ on $n_{\alpha P}'$; once a point, $T_{\delta y}$, is chosen on y, it is joined with $T_{n\alpha Pyz}$ and $T_{n\alpha Pxy}'$, and the traces of δ on yz and xy are obtained; δ is orthogonal to α.

The line common to δ and α passes through Q and is orthogonal to $n_{\alpha P}$; a line of α is always orthogonal to $n_{\alpha P}$, with respect to which it can be incident if it contains Q or simply orthogonal skewed.

In Fig. 3.7b, the plane $\alpha(t_{\alpha zx}, T_{\alpha x}, t_{\alpha xy}')$ is referred to the triplet O, x, y′, z, with $u_y' = u'$; the orthogonal n (n′, n_{xy}') to α conducted through O requires the O-1 orthogonal to the $t_{\alpha zx}$, the O-$T_{\alpha y}$* = O-$T_{\alpha y}'$ orthogonal to the O-1.

At p* = $T_{\alpha y}$*-1, the true dimension of 1-$T_{\alpha y}'$ which is the projection p′ of p, the line with the maximum slope of α with respect to π = xz, we lead for O orthogonally to π* the n*.

The corresponding n′, projection of the required orthogonal is defined, after the 2-3 parallel to the $T_{\alpha y}$*-$T_{\alpha y}'$, by O-3; 3 is the projection of the foot of the orthogonal n on α; 5 is the projection of the projection on xy; O-5 is the n_{xy}'; p and n belong to the same plane orthogonal to α, having on π = xz the trace $p_{zx} = n_{zx}$.

Reflect on the geometric properties of the elements gradually considered; we recognize the n, straight by O, which in O collects the three traces of the n; Many other exercises can be hypothesized as a corollary to this.

For example, after reversing n, a given segment, the distance between α and a plane parallel to it, is reported there; or, given the segment, height of a cone, circular right, the extreme vertex is to be the center of the bundle of generators that connect it with the points of a circumference of α of center 3 and a predetermined radius.

In Fig. 3.7c, several problems are described; the view from below is adopted; the reference triad is O, x, y′, z, with xz = π and $u_y' = u$; y′ grows from left to right towards the observer; in reality, y is orthogonal in O to sheet xz taken as a painting.

Previously, the line orthogonal to the plane a was constructed. Now, the line r is given, and we are asked to construct first a plane A orthogonal to r, then to derive R, the intersection between r and α.

We can also assign a point P for which to lead the plane δ parallel to α, that is, orthogonal to r; we ask for the intersection Q with r; then we obtain the distance P-Q, which is on δ and represents the distance of P from r, which we will try to solve in a more direct way. Clear, simple, preferable.

In the search for R, we consider for r a plane β that meets α according to s; moreover, given P in space, outside α, to obtain the traces of δ parallel to the homonymous ones of α, we consider the horizontal for P, imposing on it the belonging to δ.

The distance between the points P and Q is the hypotenuse relative to two cathetes; it can be the difference in height relative to xy, shown orthogonally to one of the extremes of the joining the projections of the points on xy.

The same measure is obtained if we operate on the difference of protrusion relative to xz, i.e., according to y′, shown orthogonally to one of the extremes of the joining projections P′, Q′.

The punctual description begins with the observation of the r, a generic line, of which the r′, the r_{zx}, the r_{xy}', the T_{rxy}', the T_{ryz}' are given; the T_{rzx} is off-screen, so we do without the r_{yz}'; the T_{ryz}' is derived from 3, common to y′ and r_{xy}'.

It is necessary to recall a fundamental concept; the plane δ, projecting orthogonally the line n_α, orthogonal to α, on any other plane β, determines on β a line n_{ab}' orthogonal to the trace $t_{\alpha b}'$ of α on β.

The same plane δ determines on α the line p_{mp} with the maximum slope of α with respect to β, and on β we obtain the orthogonal projection $p_{mp\beta}'$, which is orthogonal to the $t_{\alpha\beta}'$ and superimposed on the $n_{\alpha\beta}'$.

The real orthogonality of the projection $n_{\alpha\beta}' = p_{mp\beta}'$ with the $t_{\alpha\beta}'$ is verified between the $n_{\alpha\beta}* = p_{mp\beta}*$ and the $t_{\alpha\beta}*$, after reversing α on π by rotation around the tap track of α on π.

That said, the spatial idea of the triplet with a view from below and of the line r is recovered with attention and reflection; we proceed to overturn, in $r_{xy}*$ on xz = p, of the r_{xy} for the T_{rxy}'-3 section, obtaining both $T_{rxy}*$ from T_{rxy}' through T_{rxy}'-1 parallel to y′ and 1-2 parallel to y* = z, and of the 3 bound to T_{ryz}', bringing O-3 back to O-4.

From a $T_{\alpha x}$ point of x the $t_{\alpha xy}*$ is conducted; from 5 we get 6, being 5-6 parallel to 3-4 and 2-T_{rxy}'; we get $T_{\alpha x}$-6 = $t_{\alpha xy}'$; on xz = π, for what has been said before, the $t_{\alpha zx}$ is orthogonal to r_{zx} conducted for $T_{\alpha z}$.

To determine R′, the projection of the point R of intersection between r and α, any plane β is considered for the line r; chosen for T_{rxy}', the $t_{\beta xy}' = T_{rxy}'$-7, the $t_{\beta zx}$ is obtained by joining, by means of homothetic triangles, visible in the figure, the 7 with the out-of-range trace T_{rzx}, the point of concurrence between r′ and r_{zx}.

The points common to the homonymous traces of α and β are joined by s′ and R′ is located on r′; this point may not be required, but, given a point P(P′, P_{xy}', P_{zx}), it may be asked to draw for P the plane δ orthogonal to r.

Note that 8-P_{xy}' is parallel to y′, cos+ as well as P′-P_{zx}; consider the horizontal for P, i.e., the parallel to xy; the parallels to the $t_{\alpha xy}'$ conducted for P_{xy}' and for P′, and from 8, the parallel to z determines track 10; The 10-11 parallel to $t_{\alpha zx}$ is the $t_{\delta zx}$, and the parallel for 11 to the $t_{\alpha xy}'$ is the $t_{\delta xy}'$.

Furthermore, if the distance between P and r is required, it is necessary to obtain Q, the intersection between r and the plane δ, already conducted through P and orthogonal to r; the plane b is again used; joining the intersections 12,13, between the homonymous traces of δ and β, we obtain Q′ on r′ and Q_{xy}' through Q′-14 parallel to $t_{\delta xy}'$, 14-15 parallel to z, and 15-Q_{xy}' parallel to $t_{\delta xy}'$.

The determination of the distance P-Q is obtained by tracing for Q′ the Q′-16 parallel to P_{xy}'-Q_{xy}' and by plotting in Q_{xy}' the difference in altitude P′-16 in Q_{xy}'-17, orthogonally to the P_{xy}'-Q_{xy}'; P_{xy}'-17 is the distance P-r.

Or, the distance P-Q can be obtained by tracing for P′ the P′-18 parallel to the P_{zx}-Q_{zx}, identifying on the Q_{zx}-Q′ in 18-Q′ the difference of protrusion according to y′; by reporting 18-Q′ in Q_{zx}-19 orthogonally to the P_{zx}-Q_{zx}, in P_{zx}-19 we find the measure P-r.

Among the ways to solve the distance between point P and line r, the one described below deserves more attention, which consists of considering the plane τo that the point and the line belong to.

With the reversal of the plane, i.e., of the line and the point, on π = xz, being $u_y' = u$, by rotation of the plane around its track on xz, the required distance is that between the reversals of the line and the point.

In Fig. 3.7d the triplet x, y′, z, O is given, with $u_y' = u$, in view from below; of the line r are assigned the r′ and the tracks T_{rxy}', T_{ryz}', T_{rzx}; of the point P, which does not belong to r, the projection P′ and the projection of its projection P_{xy}' are given.

Once points 1 and 2 have been obtained, the $1\text{-}2 = r_{xy}'$ defines the T_{rxy}'; conducted for P′ the s′ parallel to r′ and for Pp′ the s_{xy}' parallel to r_{xy}', we obtain the T_{sxy}', the $T_{sxy}'\text{-}T_{rxy}'\text{-}T_{\alpha y}'\text{-}T_{\alpha x} = t_{\alpha xy}'$, the $T_{\alpha x}\text{-}T_{rzx}\text{-}T_{\alpha z}$ plot on $xz = \pi$, $T_{\alpha z}\text{-}T_{ryz}'\text{-}T_{\alpha y}' = t_{\alpha yz}'$.

To flip on xz rotating around $t_{\alpha zx}$, we have the sequence: O-3 orthogonal to $T_{\alpha zx}$, O-$T_{\alpha y}'$ returns to O-($T_{\alpha y}$) orthogonal to 3-O, 3-($T_{\alpha y}$) returns to O-3 in 3-$T_{\alpha y}$*, $T_{\alpha y}$*, off-screen, joins with $T_{\alpha x}$, obtaining $t_{\alpha xy}$* overturned of $t_{\alpha xy}'$ and with $T_{\alpha z}$, obtaining $t_{\alpha xz}$* overturned of $t_{\alpha yz}'$; homothetic triangles (not marked) are used, i.e., similarity.

Reversing the traces, we derive on the $t_{\alpha yz}$* the T_{ryz}* by similarity, considering for the T_{ryz}' the parallel to the T_{ay}'-T_{ay}*; the T_{rzx} is joined with T_{ryz}* obtaining the r*; at this point, for the 4, intersection of the s′ with the $t_{\alpha zx}$, we should draw the s* parallel to the r* and finally derive the distance P-r as the distance of the two parallel lines.

The distance that should be obtained is, instead, provided by the segment 4-5 orthogonal to the r*; it is believed that the path followed is much more rational than the previous one, and in the geometry of drawing this principle must be pursued.

On the contrary, doubt and the search for the existence of more suitable ways of solving it is necessary to confirm the continuous and generalized use of reasoning and reflection.

In Fig. 3.8, the translation of a plane figure σ of a given segment orthogonal to the plane of s is illustrated; the reference triplet is given by O, x′, y, z, with $u_x' = u$.

Of the plane α are given by the traces $t_{\alpha xy}'$, $t_{\alpha xz}'$, $t_{\alpha yz}$, on the coordinated planes and $T_{\alpha x}'$, $T_{\alpha y}$, $T_{\alpha z}$, on the axes; to assign the figure σ on α it is necessary to flip α to $yz = \pi$, place the σ_α*, a circle, and obtain the projection σ_α', an ellipse.

To proceed with the translation orthogonal to α, it is necessary to obtain the projection n′ of the line n, after determining the n*; given the translation segment on n*, the corresponding segment on n′ is obtained.

It must be specified how to read the trio; you want a view from below and from behind; the stroke of x, O-T_{ax}', in true magnitude, in the reconstructed real space, reads orthogonal to the field in O, and the extreme Tax is located behind the sheet yz.

The overturning of one plane on another is determined by the rotation around the track common to the two floors; In the case at hand, the axis of rotation is $T_{\alpha yz}$; the auxiliary plane, orthogonal to $t_{\alpha yz}$, contains O-$T_{\alpha x}$ and O-1.

The distance $T_{\alpha x}$-1, the radius of rotation of the extreme $T_{\alpha x}$, is the hypotenuse of the right triangle in O, defined by the cathet O-1 and the cathet O-2 = O-$T_{\alpha x}'$; rotating 1-2, around 1, flips $T_{\alpha x}'$ to $T_{\alpha x}$*.

An oblique affine homological relation has been determined having the $t_{\alpha yz}$ as its axis of homology; the joining of the corresponding points $T_{\alpha x}'$, $T_{\alpha x}$*, represents the center of improper homology.

The O-1-$T_{\alpha x}$* is orthogonal to the $T_{\alpha yz}$ axis; the line of the $T_{\alpha x}'$-$T_{\alpha x}$* center is oblique, so that the homology of reversal is oblique affine; the counterpart of the $t_{\alpha xy}'$ is the $t_{\alpha xy}$*; The counterpart of the $T_{\alpha xz}'$ is the $T_{\alpha xz}$*.

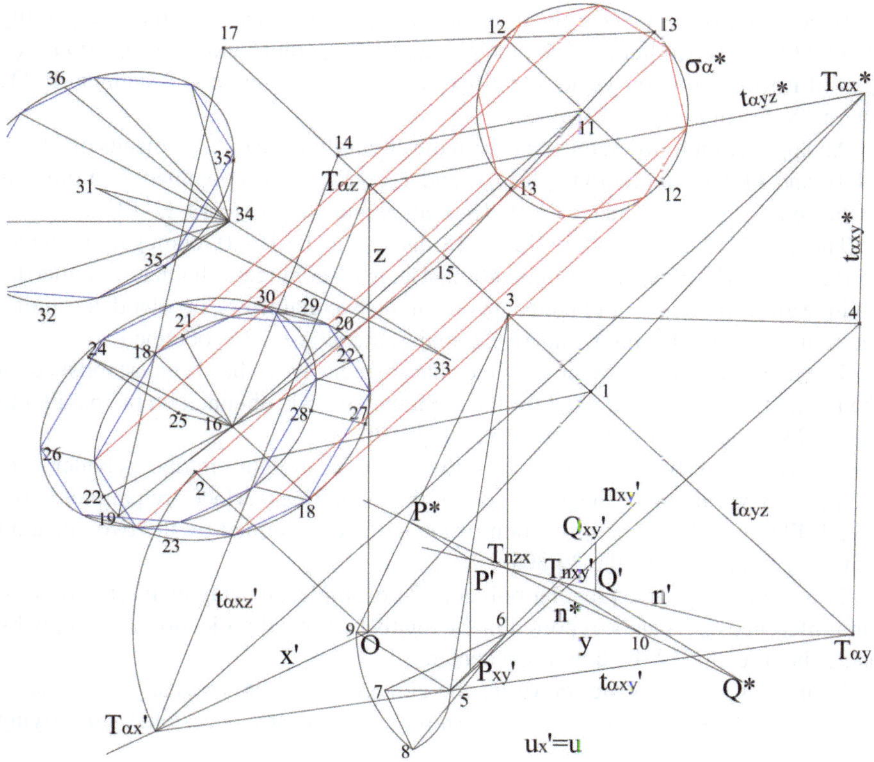

Fig. 3.8 Oblique affine homology on planes: Ellipse construction via conjugate diameters from homological data

Having chosen a 3-4 orthogonal in 4 to the $t_{\alpha xy}$*, the homologue of the 4 is the 5 on the $t_{\alpha xy}'$, being 4-5 parallel to the $T_{\alpha x}'$-$T_{\alpha x}$*; the 5 is on xy, the 3 is on yz, the 5-3 is on α and is the line of maximum slope with respect to xy; the projection of its projection on xy is 5-6.

The plane 5-6-3 has track 6-3 on y-z = π and also contains the line n orthogonal to α; this plane is reversed on yz, rotating around 3-6, after returning the 6-5, which is broken down into 5-7 parallel to y and 6-7 parallel to x′, both cathets in true magnitude.

The hypotenuse 6-8, the true dimension of 6-5, revolving around 6, overturns into 6-9; 9-3 represents the overturning of the line of maximum slope; the coplanar, orthogonal at its point P*, is the n*, which intersects the 9-6 in 10.

Parallel to 9-5 from P*, P′ is obtained on 5-3; followed by P_{xy}' on 5-6 = n_{xy}'; on 3-6 the n* intercepts the T_{nzx}; the P′-T_{nzx} is the n′, which intersects the n_{xy}' in T_{nxy}', which can be obtained with the parallel to the 9-5 conducted for the 10.

If we want to report on n a given segment P-Q, we consider on n* the true quantity P*-Q*; Q*-Q′ is parallel to P*-P′; the projection P′-Q′ on n′ and the projection P_{xy}'-Q_{xy}' of its projection on xy are identified; conversely, given on n′ a segment P′-Q′, the real value in P*-Q* is obtained.

An application follows; on α* is given the circumference σ_α*; by means of 11-14, 14-16 and 11-16, parallel to $t_{\alpha yz}$*, $t_{\alpha yz}'$, and $T_{\alpha x}$*-$T_{\alpha x}$*, 16 is obtained, homologous to the center 11; the diameter 18-16-18 is equal and parallel to 13-11-13.

The 13-11-13-15-16 and 13-12-17-18 derive the 19-16-20, conjugate diameter of the 18-16-18; Note the aforementioned pair of conjugate diameters, either by oblique affine homology implanted on one of the two diameters, or after determining the semi-axes 16-21, minor, and 16-22, major, the ellipse 23 is obtained.

The intention is to move the 23. Once the segment 16-24 has been established on the parallel to the n* conducted for the 16th, the 16-25 is obtained on the parallel to the n′; 25 is the center of the translated ellipse 26.

The 16-25 can be the projection of the axis of a straight circular cylinder with the base 23 resting on α; the generators of the cylindrical surface are parallel to the 16-25, like the 27-28; the projection of the surface is defined by the two apparent contour generators, like the 29-30.

The vertices of a given regular polygon, for example, an octagon, inscribed in the circumference σ_α*, can be reported in the relative elliptical projection 23, simply by using the direction of the homology center.

In any case, to construct or verify the correctness of the results, one can draw on the rules of symmetry, centrality, parallelism, and any occasional simplifying property.

By translating the projection of the octagon from 23 of center 16 to 26, of center 25, the relative joinants are the projections of the edges of the regular prismatic volume right with the octagonal base belonging to α.

Move the segment 31-33, parallel to n*, according to the 33-34, parallel to P′-P*, to 31-34, parallel to n′, according to 33-34, parallel to P′-P*.

Point 34 is the projection of the vertex of the right circular cone, having the base equal to σα* and the height equal to 31-33; the two generators 34-35 are those of apparent contour; 34-36 is a generic generator of the surface of the cone.

The centers 16 of the base of the cylinder and 31 of the base of the cone are coplanar, on α; the height of the cone develops in the opposite direction to that of the cylinder, remaining on opposite sides with respect to α.

By translation, the projection of the regular octagon, already known, is inscribed in 32; its vertices, connected to vertex 34, complete the image of a regular straight pyramid.

Whoever observes the space defined by the triad according to the view from below sees Tax moving away from O; then, Q moves away on n with respect to P, as Tax with respect to O.

It is concluded that the cylinder develops in height as it approaches and is visible, as is the octagonal prism inscribed in it; the cone and the pyramid inscribed in it develop in reverse; the top moves away; the aforementioned volumes are not in sight.

Let it be reflected; on the contrary, it is recommended to think independently, in search of one's own conviction; if this, supported by the definitive absence of doubts, should disagree with what has been concluded above, let no victory be sung; Recognize that everyone can err, even the writer.

The scholar is left with the problem of identifying which and for what part of them are the elements in view in the images of the four volumes. We also invite you to consider the interpretative hypothesis of a view not from below but from above, in order to highlight, even if only mentally, what changes in the results acquired.

In Fig. 3.9, we deal with the problem of the carry-over of a rotating segment with increasing dimension according to an Archimedean spiral, belonging to the yz plane in a reference O, x, y, z′, with $u_z' = u$.

Once the plane yz has been reversed on xy = π, by rotation around y, on yz* the spiral of origin A* is drawn, assimilating the successive arcs to segments-chords, of increasing radius in constant measure and angle at the constant center; one is followed

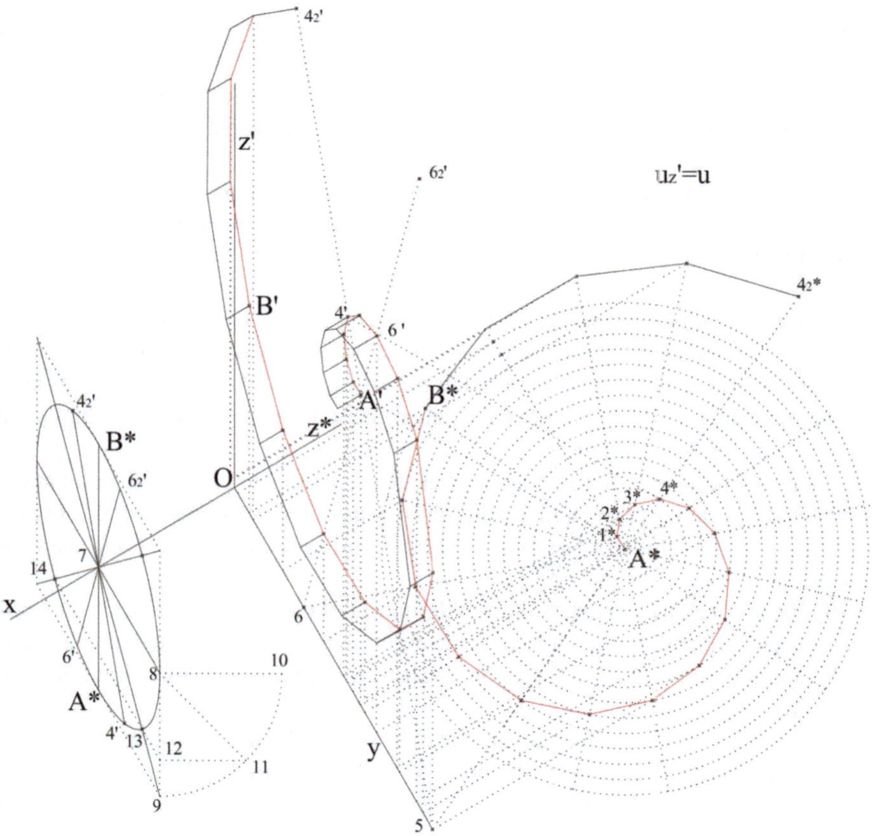

Fig. 3.9 Carry-over of a rotating segment with increasing dimension according to an Archimedean spiral, belonging to the yz plane in a reference O, x, y, z′, with $u_z' = u$

by the other with a positive increase, also increasing compared to the previous one, but not proportionally.

The carry-over of each vertex in projection on yz′ is obtained by considering from the point the parallel to $z^* = x$ up to y and bringing this distance parallel to z′, being $u_z' = u$; between the two planes, an oblique affine homology of the y-axis can be recognized.

Observe the resulting segments; they are strongly elongated or shortened if the direction of the projection is close to the long or short diagonal of the rhombus, projection of the square belonging to yz with the sides on y and z, and vertex in O.

Let it be reflected; the spiral has made a complete turn passing from A* to B* and in projection from A′ to B′; the distance A*-B* is the pitch of the real and constant helix but of rotating radial direction; a second turn of the spiral is mechanically reproduced by returning the pitch value unchanged in the next radial direction, according to the right-hand direction adopted.

In projection, the pitch is variable; in the second round, it is constant and equal to A*-B*; for each vertex, one can operate, applying the oblique affine homology, defined by the y-axis and the pair of homologous points A*, A′, since A*-A′ is the direction of the improper center.

For example, from 4* we get 4′ with 4*-4′-5-A′-4′ and with 4*-4′ parallel to A*-A′; however, the top of the first round is needed; in the second round, the solution may be different.

There is no need for the availability of overturned top management; note that the distance between a given inverted vertex of the first turn and on the same radius of center A*, the corresponding vertex of the second turn is constant and equal to A*-B*.

It is, for example, 4*-42* = 6*-62* = A*-B*; the 4_2*-4*-A* direction has the 5-A′-4′-4_2′ as its homologation; the 4′-42′ is, in projection, the shortening or lengthening of A*-B* in the 5-A′ direction; the 6_2*(off-screen)-6*-A* direction has as its counterpart the 6-A′-6′-6_2′; the 6′-6_2′ is, in projection, the shortening or lengthening of A*-B* in the 6-A′ direction.

Then let 7 be a point of x and on the plane for 7, parallel to yz, consider a square of side A*-B*, which in projection is a rhombus; consider the diagonals, which are axial directions for the ellipse projection of the circumference of center 7 and diameter A*-B*.

By means of an in-situ reversal, from the end 11 of the diagonal we reach the semi-major axes 7-13 and minor 7-14; once the ellipse has been constructed, the diameter 4′-4_2′, parallel to the 5-A′-4′ direction, must be brought back to this direction in 4′, obtaining 4_2′.

In the same way, even if we do not have the 6_2* on the field, to obtain the 6_2′, the radial direction A*-6 relative to the point 62* of the inverted spiral is sufficient; the direction of the 6-A′-6′ homologation identifies the diameter 6′-6_2′ on the ellipse, which is transferred to the homologation in 6′ identifying the 6_2′.

Reference to ellipse diameters allows you to obtain any point in the spiral of any revolution following the first, but available as a continuous curve and not a polygonal one for the first revolution.

The translation according to x transforms the spiral line into a spiral strip with a constant width, or, if you like, with a variable width, in any case, for example, with a fixed constant increase, developing a helical line on a spiral base, as well as, if required, a helical surface; just rely on your imagination.

Regarding the overturning of the quarter of a circumference, for correctness of reference to the axis of the overturning, this should have used the axis A*-7.

It follows that 8-9 is identified with A*-7, the arc has a center in 7 and begins in A*; from the consequent extreme of the 45° diagonal with A*-7 conducted through the 7 and common to the arc, the horizontal is considered.

At its point of intersection with A*-7, the parallel is led to the side of the rhombus that contains A*; with points 13 and 14 on the diagonals of the rhombus, it defines and the semi-axes of the ellipse, 7-13 and 7-14.

Reflect on the possibility and convenience of the many reversals, provided that the results are correct; among the many, for example, the line B*-8 can be chosen as the homological axis of overturning.

The lawfulness is ascertained and a pseudo-geometric procedure is accepted, exceptionally, provided that nothing changes in the outcomes, without, therefore, affecting the rigor of geometry.

Translation and Rotation

In Fig. 3.10, we deal with the problem of translation parallel to z and simultaneous rotation of one-sixteenth of an angle around the z-axis, applying it to a square of xy.

In the space, there is a volume articulated in overlapping segments, whose module, between two planes parallel to xy, is delimited by four crooked surfaces, each being part of a hyperbolic paraboloid.

The circumference c, traced on xy of the vertical cylinder of the z-axis, circumscribed by squares, has center in O and has the diameter A-O-B; from c*, by homology of the y-axis, pair of homologous points D′ and D* and improper center D′-D*, c′ is obtained.

From the subdivision of c* into sixteenths, the homologous subdivision on c′ is obtained; The quaterns of points, extremes of the diagonal orthogonal diameters of the rotating squares, become the extremes of the projection of each square.

The initial square is not 1-2-3-4, but 7-8-9-10; 5-6 is the sixteenth of the step relative to the rotation of a complete revolution; each vertex describes a helix on the cylinder; The perimeter polygonal of each square joins the extremes of a pair of conjugated, diagonal diameters of the quadrilateral.

The quadrilateral, having diameters 1-3 and 2-4 as diagonals, is a rhombus; the quadrilateral, having diagonals A-B and C′-D′ is a rectangle.

The points joining the points of the subdivision into 16 equal parts of the side 7-8 and its upper opposite constitute a non-developable striped surface, a portion of a hyperbolic paraboloid; the apparent outline appears uneven; the overall view is plastic due to the succession of crooked surfaces.

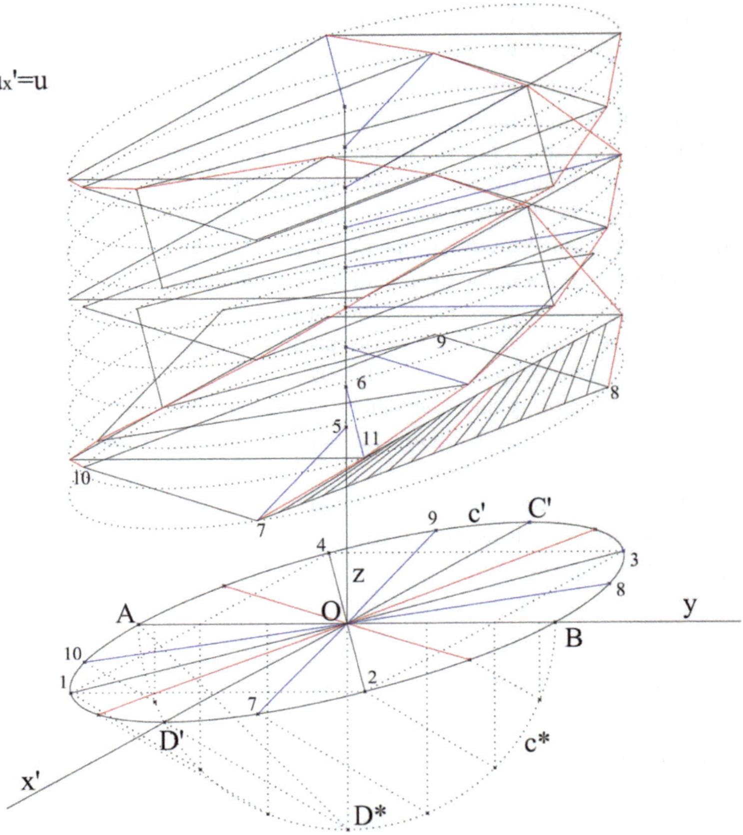

Fig. 3.10 Translation parallel to z and simultaneous rotation of one-sixteenth of an angle around the z axis, applying it to a square of xy: the initial square is 7-8-9-10

It requires a great deal of attention and indispensable reflection, essentially formative, to identify the junctions of each of the vertices of the initial quadrilateral with the corresponding one, following the translation and rotation of the projection of the previous square.

These are four polygonal humps of straight segments, whose vertices are points of rototranslated cylindrical helices; in space, they are all segments equal to each other; their length develops in a different direction but with a constant slope.

In projection, precisely because of the aforementioned geometric properties, they differ so much that it is unrecognizable for each of them to have the spatial support of a single cylindrical helix in rototranslation.

The semidiagonals are four equal radii, but the appearance cannot confirm this, without reference to the corresponding semi-diameters in the basic matrix ellipse c′, a projection of a circle c of xy.

To provide a reading guide, for each position assumed by the square in rotation of one-sixteenth of the pitch and one-sixteenth of the rotation angle, one of the radii has been highlighted, for example, the 5-7 and the subsequent 6-11, crooked with each other.

The exercise can still be surprising, if it is not a square but a regular polygon of n sides in a rototranslation, which uses an nth of a turn angle and an nth of a pitch.

The helices will be better perceived, reducing the surfaces of the hyperbolic paraboloid quadric to surface ripples; imagination, curiosity, challenge, and maturing the study to the conscious possession of unique and innovative design skills.

The image of the exercise is somewhat incomplete, to be imagined, to be thought, to be solved; the conceptual implications must prevail over the operational ones; delegating to the subsequent and personal completion, it is reiterated that in order to provide an image, it is a priority to understand, deepen, get involved not by the how but by the why.

An interesting application of rototranslation of incident orthogonal lines in oblique axonometry is the representation (Fig. 3.11) of a helical staircase on a straight circular cylinder.

It was decided to draw not a staircase, but that staircase, of which the dimensions useful for the representation in oblique axonometry on the xy picture are given; moreover, the relative drawing must be "instructed" on sheet A.B.C.D = 70 × 50.

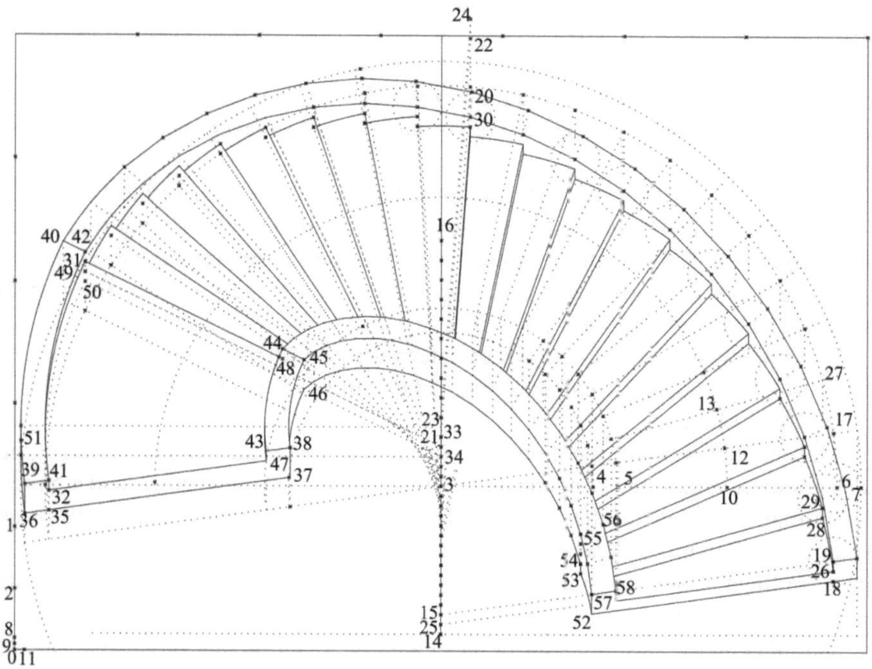

Fig. 3.11 Spiral staircase projection: Axonometric development of a helical stair with steps and landings

The steps are shown on a helical slab; they are all the same, they have a rectangular riser and a circular sector tread; the slope of the helicoid is that of the step; it is given by the ratio between the riser and the advancement of the tread, measured on the average axial circumference of the stair width.

The slope of a straight flight of stairs, with a rectangular tread, on any line parallel to the parallel edges, is constant; on the other hand, that of the helical ramp is smaller on a propeller with a larger radius and vice versa.

The staircase has, among several functions, that of allowing the overcoming of a difference in height; the one between two floors is called interfloor; the lower plane can be horizontal or inclined, and can be parallel, or not, to the upper plane of arrival.

The tread can also have a slope, which can be in agreement or discordant with that of the ramp supporting the steps; The riser corresponding to the slope of the tread is added to or subtracted from the actual riser and the resulting value in relation to the horizontal displacement between the edge of one tread and that of the next tread determines the slope of the stair.

The projection of the outer edge of the tread on the horizontal plane may not coincide with that of the inner edge of the tread below. What matters for the slope of the staircase is the vertical displacement compared to the horizontal displacement.

The step of the aerial ladder can be supported by an axial beam or be wedged bilaterally or on one side only to a flat or cylindrical wall, to a pillar, column, or suspended from tie rods.

The step can be cantilevered and can show the thickness tapered towards the free end; the riser can be virtual. The step can have virtual tread and riser, transformed into a cylindrical peg in metal, wood, or other material.

The ladder with a light structure and limited length can be made of wood, iron, aluminum; The rigid type can be fixed or mobile, to be placed with an appropriate slope.

A ladder can be a scissor ladder of limited length; it can be of variable length for the assembly of additional sections added by interlocking or by sliding ropes, with the uprights parallel or with a small upward convergence, rigid or flexible if made of rope; the treads are narrow and orthogonal to the axis of the staircase; The vertical staircase has an infinite slope.

These are functional ladders for particular purposes; Also mentioned are the access ladders to ships or aircraft, equipped with safety railings, reticular, and with handrails, retractable or liftable.

Masonry stairs perform not only the function of connecting points in height, but also aesthetic values involving the use of precious materials and artistic and architectural ornaments; they can have the development of straight ramps with intermediate and extreme landings, as well as parapets with balustrades and molded cornices or metal railings of the most varied expressive shapes of times, places, environments, societies, civilizations, etc.

The planimetric development in the inter-storey can be one ramp, two side by side in the opposite direction, according to a rectangle, the trace of a parallelepiped stairwell, with several ramps, whose axis repeats the polygonal trace of the room that contains it, or of the structure that surrounds and from which it jumps.

In the corners, non-rectangular treads can be used, i.e., polygonal landings can be inserted. Depending on the size of the stairwell in relation to the width and size of the stairwells, there can be stairwells, often colonnaded, or load-bearing spine walls.

The planimetric development can be forked or combed, simple or double opposed, in search of unthinkable, effective, very perceptive scenographic effects.

In historical and architectural times, the art lavished on the design and construction of many types of staircases with original schemes, testimonies of bold and innovative structures, has often exceeded the limits of function and sobriety, with ostentation of wealth, pomp, waste, and indelible manifestations of unforgivable economic, social, political, and religious arrogance.

The scales, which present appreciable singularities, are of unspecifiable number, unimaginable. The horizontal projection of the axis of a staircase or its ramp can develop according to straight or curved sections, which can be circular, simple, compound, or S-shaped, elliptical, oval, cycloid, involute, etc.

In wells, cylindrical towers, large tanks, cross-linked piles, chimneys, antennas, bridge pylons, masts of ancient vessels, to access roofs or large roofs, in many industrial structures, ladders can be vertical protected by horizontal ring railings, which can share the function of handrails with the same rungs of the ladder.

In the cavity of the domes, it is possible to obtain helical staircases with variable radius and inclined free heights, and sacrificed for the user, even for the limited width.

In some castles, the staircases, with a helical development inside a cylindrical tower outside the structure, have single steps interlocking with each other, and their lower part constitutes a continuous form of the intrados with a flying vault, with a circular arch on an axial plane; There are even more complex solutions.

We can speak of ladders in the climb to the mountain or in the speleological descent, we can speak in a metaphorical sense of climbing to success, we can recognize the direction of the development of the ladder dug into the pier for the mooring of the boat.

One can think of the cordon in the English garden or in the park, the bleachers in the theater or in the arena or in the stadium, the stepped roof, of constant or variable slope.

For the safety of walking a staircase, it is necessary that the value of the slope is constant; In supermarket escalators, the value of the slope varies from zero to a maximum approximately equal to that currently adopted.

It should be noted that the user standing on a single step does not feel the variation in slope between two successive steps; each step, reaching each position of the trajectory of the race, always assumes the slope structurally envisaged with the adjacent steps.

It is also appropriate to reflect on the handrail; in very steep or vertical stairs, the handrail consists of the steps above, on which the hands rest uphill and downhill; particular attention should be paid to the handrail of the escalator.

And again, we can think of the stairs, whose intrados is not a flat or helical ramp, but reproduces, in search of an aesthetic effect, the stepped split characterizing the upper part.

The scholar is suggested to make an effort to recall other possible experiences of scales to be treated; curiosity and imagination can take you a long way. A frequent experience is the escalator.

In the helical cylindrical staircase, at the completion of the first 360° turn, the difference in height overcome is the pitch of the propeller; it can be equal to the inter-storey to be exceeded, or to a multiple or sub-multiple, whole or fractional of it.

Taking into account that between the extremes of the difference in height to be overcome, there must be a whole number of risers, the ratio between the values of the resulting riser and the tread proportionate to it must be close to the functional one.

The pitch, according to the vertical between the lower helical surface, which is the visible intrados of the s-thick slab of the aerial ramp after a complete turn, is constrained, at least, to respect the minimum free span useful for the use of the staircase.

The external helical metal ladders on a modest radius cylinder, commonly called spiral ladders, have circular sector treads. The slope of the propeller line varies from the minimum on the perimeter propeller to the maximum on the propeller near the central load-bearing shaft, of a value close to infinity.

The real travel propeller, depending on the radius-distance from the axis, can generally be considered to be half the width, if this is intended for the passage of only one user at a time.

If you want to allow the passage of two users side by side, the adequately increased width does not solve the safety problem; the user closest to the axis faces a greater slope, the rise being constant, and the tread reduced.

Having to guarantee the safety of the most disadvantaged, the canonical slope adopted according to the formula $p + 2a = 64$ must be verified along the propeller of travel, at least half of the width of the potential user, added to a prudential value of pleasantness for a serene use of the staircase.

Having chosen (Fig. 3.11) to represent a helical staircase in oblique axonometry on the xy and $u_z' = u$ frames, use a 70 × 50 cm sheet; on the monitor is 0-1 = 10 cm.

A scale of 1:10 is adopted on the sheet; it follows that 0-1 represents 100 cm, 0-2 is worth 50 cm, 0-8 is worth 10 cm, 0-9 is worth 5 cm, 3-7 is worth 345 cm; the diameter of the outer cylinder is 690 cm and the inner one, the 3-4, is worth 300 cm; the width of the staircase, including the thickness of the skirting walls, 4-7, is 195 cm, the thickness of the wall 4-5 = 6-7, is 7.5 cm; the net width of the flight of scale, 5-6, is worth 180 cm and the radius of the axial circumference, 3-10, is worth 247.5 cm.

The 0-11 = (8:10) of 0-8, equivalent to 8 cm, is reported four times consecutively on the circumference of radius 3-10, starting from a point of 12, constituting the arc of the tread of 32 cm in 12-13; this arch must be carried over 20 times, as there are 20 steps.

Twice 0-11 is equal to the 16 cm lift, brought back on the vertical axis in 14-15; the 320 cm spacing is given in 14-16. Before proceeding with the design of the steps, it is advisable to check the overall dimensions of the drawing, since every detail necessary in the field must actually be such.

The first lift, in plan, is 3-17; to it, for the choice $uz' = u$, corresponds the baseline 14-18 and the upper edge 15-19, being 17-19-18 vertically aligned.

The riser, represented in the plan by 3-20, corresponds to the lower line 21-22 and the edge 23-24; the drawing already appears to encroach on the outside sheet; A $u_z' = u:2$ must be used.

For this choice, the 14-15 is halved into 14-25, and the first lift is defined as 14-18-26-25; the outer arc of the first tread 17-27 of radius 3-17 is returned with the center in 25 in 26-28, being 28 on the vertical of 27; the second lift is 25-28-29-15.

The step, whose tread has the top in 30, is the last to still have the riser in sight; the following steps, up to landing 31-32-47-48, do not show the riser, nor a part of the tread, each hidden by the edge of the next tread.

The thickness of the landing measured vertically, 31-49-50, is equal to two risers, while the thickness of the slab, which supports the steps reported and integrated, is equal to one riser.

The helical surface of the soffit is 4-7 wide; the height of the two curbs flanking the board is equal to three risers, 37-38 = 36-39; the surface of the curbs, for the horizontal flat section of the board, is delimited by four arcs of horizontal circumference.

For the external curb, the circular arches 39-40 and 41-42 are visible, under which the arch 32-31 is partially visible; for the internal curb the circular arches 43-44, 38-45 are visible, while the 47-48 is hidden; the circular edge 37-45 is partially visible; The intrados of the board is not helical, but flat and horizontal.

The helical ramp develops between two cylindrical surfaces, the circumscribed one, with radius 3-7 and the inscribed one, with radius 3-4; of the circumscribed cylindrical surface, pertaining to the landing, since the horizontal projection of the landing exceeds the horizontal diameter of center 3, only a small part is visible, defined by the curvilinear triangle 36-39-51-36, being the 51 common to arches 36-49, lower, and 39-40 upper.

The internal part consists of the circular arches, 37-46, lower, and 38-45, upper, both horizontal, which stand at the vertical sections 37-38 and 46-45; 43-44, the inside edge of the curb, is in sight; the underlying 47-48 is not completely hidden, but a curvilinear triangle is visible, practically visible only if greatly enlarged; a short arch develops from 47 to the vertical tangent to arc 43-44; The 46-45 segment, in sight, is virtual, as is the 48-44, which is hidden.

Here, it is necessary to insist on highlighting that the competence of "drawing" must consider as a priority the maximum attention, observance, and precision on the so-called minimum details, the neglect of which cannot fail to arouse suspicion on the ability and not on the will to notice it; the image must be lived, possessing every reason for it.

For the construction of the internal curb, it is advisable to proceed from the first step; note that 52-53 is an arc of horizontal circumference; 55-54 is equal to three lifts; the filling relative to 52-53-54 prevents the soffit from ending at an edge on the support surface.

In correspondence with the vertical plane of each riser is the trio 56-55-54; are the points of three cylindrical helixes; the one relating to 54 is shifted down by three risers compared to the one relating to 55; the one relating to 56 belongs to a cylinder of greater radius and has a lower slope; the propeller of the 55 begins in 57, that of the 56 in 58.

Similar considerations apply to the corresponding propellers of the external curb; the horizontal circular arch intersections of the treads on the smaller cylinder are all visible; the sections of generators common to the aforementioned cylinder and to the risers are visible only on the right.

Some reflections: the ramp develops as it goes up, but this is suggested by appearance up to a certain point, and then it seems that it continues downwards; this perceptual effect is accentuated by the lack of references on the impost plane of the ramp. Also, note that the interfloor of the staircase is reached before the step; To make the inter-storey coincide with the pitch of the helicoid, it is necessary to reduce the radius, i.e., to appropriately widen the width of the tread along the arc of average radius, 10-12, reducing the slope.

Note that in this case, the base circle is divided into 20 parts; For reasons of space, the ratio u_z':u must be reduced, without prejudice to the graphic scale adopted on xy; the central loop, which can barely be seen in the figure, appears closed when the turn is completed, and is repeated in the multi-storey staircase.

The scholar must hypothesize personal questions, to be answered; cannot limit itself to taking note of illustrated cases; by repeating them, he has not settled the accounts with their potential.

Consider, for example, whether, in an exposure from above, a partial view of the soffit is possible, or, up to what development, in relation to the slope ratio, the introduction of a landing limits the minimum functional height of the free opening; What changes if the soffit is helical with an oblique generator and not parallel to the impost plane?

Central, Axial, Orthogonal, and Oblique Symmetry; Symmetry with Respect to a Plane

Let x′, y, z be the projections of the axes (Fig. 3.12); yz is the picture, plane on which the points of space are projected in the direction $u_x' = 2u_z = 2u_y$, with a free elongation of the unit 0-3 according to x, equal to twice the unit 0-1 adopted on π.

Consider on xy the projection quadrilateral 0-4-9-7-0; don't bet that it is the image of a square, because, contrary to appearance, it is a rectangle; 0-4 represents $2u_x$ is oblique projection of 2u, and 0-7 is equal to 4u.

It follows that the truncated pyramid of height 10-12 has rectangular bases 0-4-9-7-0 and 13-15-14-16-13; the upper minor base and the side faces 4-9-14-15-4 and 7-9-14-16-7 are visible.

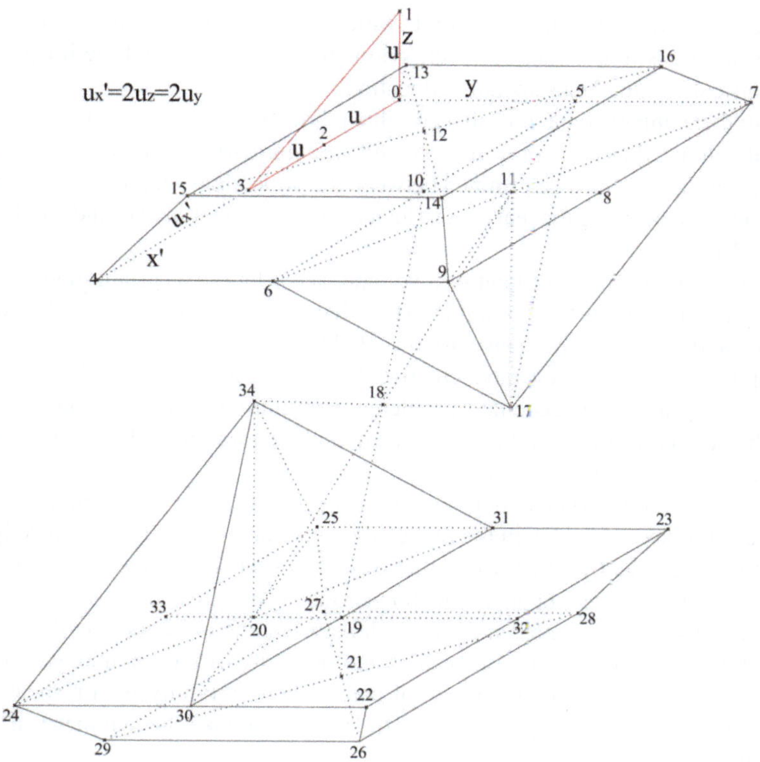

Fig. 3.12 Given two figures that do not appear to be symmetrically related to each other, it is possible to identify the center of symmetry at point 18. Plane's central symmetry is thus determined between the projections, defined with respect to the projection, on xy, of the real center

On the level of 0-4-9-7-0, consider the quadrilateral 5-6-9-7-5; 5-7 is equal to 2u, and 5-6 is the oblique projection of 2u; it is an image of a square; of the straight pyramid of height 11-17, the faces 6-9-17-6 and 7-9-17-7 are visible, identical to each other.

Once the fundamental reading problem has been solved, let a center 18 of symmetry be chosen; 10-18 = 18-19 and 11-18 = 18-20; the symmetrical 19-20 of the 10-11 retains the size and direction, while the direction of reading is reversed; the same applies to 19-21 and 20-34 compared to 10-12 and 11-17, in the symmetry of center 18.

In the central symmetry of the figures belonging to xy, or parallel, or flat in any case, it should be noted that the positions, the linear and angular dimensions are preserved, but what is on the right will appear on the left, what is above will be at the bottom; the faces that are visible are no longer visible, and those that were hidden appear.

The center of symmetry is a point, which is not defined by the projection 18 of its spatial position alone; however, symmetry takes place, and the image of the symmetrical volume lacks spatial collocation.

With the complete data, it is also possible to identify the projection on xy of both the real volume, the center of symmetry, and the symmetrical virtual volume; a plane central symmetry is thus determined between the projections, defined with respect to the projection, on xy, of the real center; let it be reflected: symmetry binds projections on any deposit.

Think again of the movement of a volume, according to a certain trajectory, and of its symmetry, at the sight of a thick shell that presents a concavity, polyhedral or quadric, and of the relative convexity revealed by symmetry, etc.

Inspiration, imagination, and creative freedom, can lead the scholar to pose singular, original, unprecedented problems, committing him to reasoning, reflection, the conquest of a desirable self-esteem, and the awareness that his preparation depends on personal resourcefulness.

Finally, it is appropriate to point out, without equivocation, that symmetry cannot and must not be confused with the reflected image; the latter is the acknowledgment of a presence, absent in the former, that is, of an observer, with respect to whom a perspective process resting on a reflecting surface is manifested.

For axial symmetry, one can distinguish between one orthogonal to the axis and one oblique to it; nothing prevents the same obliquity of the oblique axonometric projection in use; in axial symmetry, of course, zero obliquity with respect to the axis is excluded; the setting is problematic, which can take the symmetrical image out of bounds.

The axis of symmetry is any line, generic or particular, even coplanar to a plane figure, incident or not with the given geometric entity; this can be fixed or in motion with respect to the axis, according to simple or complex, concordant or discordant trajectories; imagination allows choices of perceptive effect, even relevant.

Assume (Fig. 3.13) the triplet x, y, z′, O, being $u_z' = u$, on a plane $\alpha(t_{\alpha xy}, t_{\alpha yz}, t_{\alpha zx}, T_{\alpha x}, T_{\alpha y}, T\alpha z')$, a triangle P-Q-R, a line $r(r', r_{xy}, T_{rxy}, T_{ryz}')$, orthogonal axis of symmetry; the symmetrical triangle is required.

Without prejudice, R can be established as a point of r on α, united to the symmetric; let the plane β defined by P and r in (P_β) and (r_β) and by Q and r, in (Q_γ) and (r_γ) be reversed on xy.

The orthogonal segment between (P_β) and (r_β) and that between (Q_γ) and (r_γ), identify (P_r) on (r_β) and (Q_r) on (r_γ); from them, we arrive at the r′ to P_r' and to Q_r'.

The symmetrics of P'-P_r' and Q'-Q_r' derive P_s' and Q_s'; the triangle P_s'-Q_s'-R' is in orthogonal axial symmetry with P′-Q′-R′; the plane α_s is symmetric of α; every line R-K of α has R-K_s as its symmetric, and its improper point has the improper point of the symmetric as its symmetric.

Lines parallel to the above have lines parallel to the symmetric as symmetric; the plane for R orthogonal to r intersects α according to the line u; each point of the u has as symmetric a point of the same u, which is a line united but not of united points.

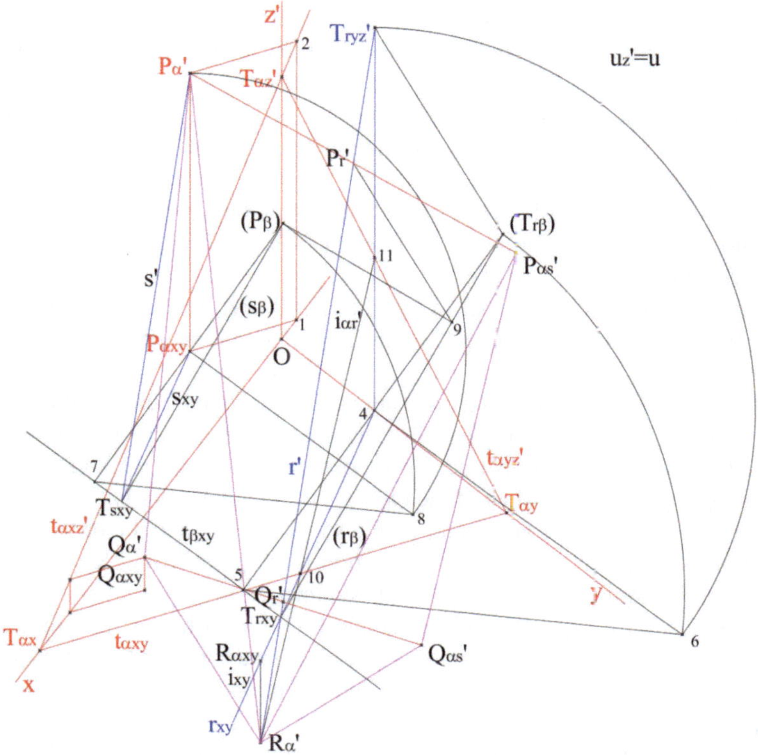

Fig. 3.13 Determination of the symmetric triangle of the given one P-Q-R with the line r(r′, r_{xy}, T_{rxy}, T_{ryz}') as orthogonal axis of symmetry

This is the line common to α and α_s; the points of you and the unit us are in central symmetry with respect to R; the symmetrics on α_s of the lines of α parallel to you have the same improper point.

The scholar can deepen the alternatives; he can carry out investigations on the diversity of results, choosing α different positions with respect to r, for example, orthogonal, parallel, which contains r; he will verify much simpler, simplified, particular, and immediate situations.

For α orthogonal to the direction of projection, at 45°, and for r parallel to α, can orthogonal axial symmetry lead to the superimposition of a figure α with symmetry? Can you have a connection?

Are there any relationships between the projections on xy of the figure and the symmetric, useful for providing different ways to obtain symmetry in space?

It is sometimes said that a study is 360° on the plane; in space, it is more than such, just as infinite, but it is capable of giving satisfactions that cannot be compared.

Now we describe the procedure followed for the derivation of the symmetric of P_α; P belongs to α; the horizontal of α for P has $P_{\alpha xy}$-1 and 2-P_α' projections parallel to $t_{\alpha xy}$.

Construct the plane β for P and r, consider for P the parallel s to r; the s_{xy} parallel for $P_{\alpha xy}$ to $r_{xy,}$ and the s′ parallel for P_α' to r′ identify the trace T_{sxy} and the T_{sxy}-T_{rxy} is the $t_{\beta xy.}$

Of β, turn r and s on $\pi = xy$, obtaining (r_β), (s_β), (P_β); to obtain (r_β), consider 4-5 orthogonal to $t_{\beta xy}$, 4-6 = 4-T_{ryz}' orthogonal to 4-5; 5-6 is transferred to 5-4 in 5-$(T_{r\beta})$; T_{rxy}-$(T_{r\beta})$ is (r_β).

To obtain (P_β), we consider, analogously, $P_{\alpha xy}$-7 orthogonal to $t_{\beta xy}$, $P_{\alpha xy}$-8 = $P_{\alpha xy}$-P_α' orthogonal to $P_{\alpha xy}$-7; the 7-8 is back on the 7-$P_{\alpha xy}$ in 7-(P_β); the T_{sxy}-(P_β) is the (s_β), parallel to the (r_β).

From (P_β) we lead orthogonal to (r_β) the (P_β)-9 and, by similarity, we consider for the 9 the 9-P_r' parallel to $(T_{r\beta})$-$T_{\rho\psi\zeta}'$; the P_r'-$P_{\alpha s}'$ is symmetrical to the P_α'-P_r'; the symmetrical of P_α' is $P_{\alpha s}'$.

Once a point Q on α has been chosen, after verifying that it belongs, the procedure just illustrated is repeated to obtain P_{as}'; the plane γ of Q and r is considered and reversed by rotating around its trace on xy, obtaining (r_γ), (s_γ), (Q_γ); Q_r' is obtained and the Q_r'-$Q_{\alpha s}'$ is symmetrical of the Q_α'-Q_r'; the symmetric of Q_α' is $Q_{\alpha s}'$.

The $i_{\alpha r}$, common to α and r and of tracks 10 on $t_{\alpha xy}$ and 11 on $t_{\alpha yz}'$, has for projections $i_{\alpha r}'$ and $i_{xy} = r_{xy}$; their intersection is the point R(R_α', $R_{\alpha xy}$); belonging to the axis of symmetry r, it is joined to its symmetric.

The projection P_α'-Q_α'-R_α' of the triangle P-Q-R of α has as its symmetric the symmetric $P_{\alpha s}'$-$Q_{\alpha s}'$-R_α' of α_s symmetric of α; the line u common to the two planes in axial symmetry contains R; it has already been mentioned; it can never be too much to reflect on the u.

The axial symmetry discussed so far is orthogonal, and the example analyzed hinges on the symmetry of a plane figure; his plane, in particular, is incident on the axis to the finite; the search must be repeated on the orthogonal and parallel planes.

A figure on a plane orthogonal to the axis has the coplanar figure as its symmetry, obtained by central symmetry; its center is the point common to the plane and the axis.

A figure on a plane parallel to the axis has a plane parallel to the axis as a symmetrical plane and contains the figure of symmetrically read edges.

Much more executive effort is required for the symmetrical of a volume; if it is a polyhedron, it can operate by symmetry of each vertex; for the symmetrical curve of a circumference, a series of results is predictable as a function of its position with respect to the axis.

The apparent outline of the symmetrical volume is different, and its predictability is the result of extreme reflection, which can benefit from experience with great prudence.

For the spherical volume, one can think, for example, of considering the bundle of planes conducted by the axis; each of them is a circumference; its diameter parallel to the axis is considered; of its extremes, the symmetrical ones are obtained; Their set constitutes the apparent outline, in an ellipse projection, of the symmetrical spherical volume.

On reflection, the scholar should pose the problem of identifying two conjugate diameters of the aforementioned ellipse; Certainly, the way is relatively simplified.

An elementary example of oblique axial symmetry is given by the semi-ellipse delimited by a diameter, from which, by symmetry, which has the aforementioned diameter as its coplanar axis, the other half can be obtained, operating with obliquity given by the diameter conjugated to the axis.

Is it legitimate to feel an expert on the subject, knowing how to solve the example suggested above?

Of course, there is doubt, and for a reassuring awareness of one's own formation, it is considered desirable to deepen the oblique axial symmetry.

It is proposed to solve the oblique axial symmetry of an octahedron, whose maximum square central section belongs to the plane α; the axis is orthogonal to α, and the direction of symmetry forms an angle φ with the axis.

Let us choose (Fig. 3.14), as a reference, the triplet x′, y, z, for which the frame π is yz; so that the example is not too simple, do not adopt the angle of 45° as the inclination of projection on π, but a smaller value, such that $u_x' = u{:}2$.

On the plane $\alpha(t_{\alpha xy}', t_{\alpha xz}', t_{\alpha yz})$, parallel to y, we want to report the maximum square section of a regular octahedron; its axis, orthogonal to α by its center, let it be

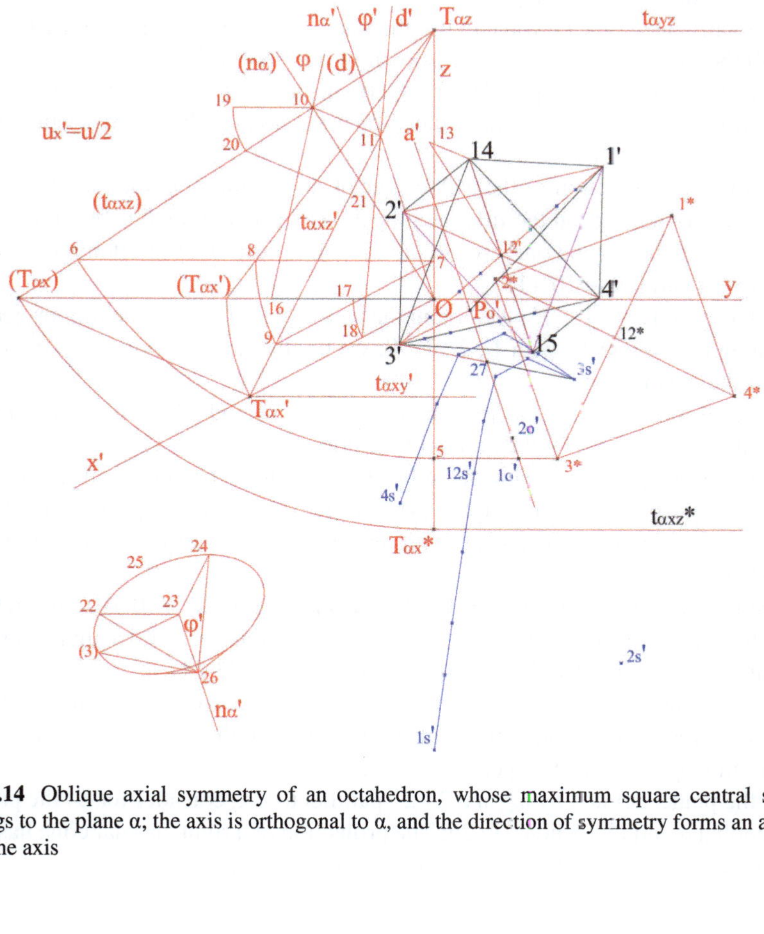

Fig. 3.14 Oblique axial symmetry of an octahedron, whose maximum square central section belongs to the plane α; the axis is orthogonal to α, and the direction of symmetry forms an angle φ with the axis

assumed to be an axis of oblique symmetry, whose direction forms the angle φ with the axis.

The aforementioned carry-over requires the overturning of α on yz by rotation around $t_{\alpha yz}$; the $T_{\alpha z}$-$T_{\alpha x}$ must be overturned after determining its true size.

Therefore, flip the O-$T_{\alpha x}'$ on y into O-($T_{\alpha x}'$), which doubles in O-($T_{\alpha x}$), being $u_x = u = 2u_x'$; the $T_{\alpha z}$-($T_{\alpha x}$) goes to $T_{\alpha z}$-$T_{\alpha x}$*; the $t_{\alpha xy}'$ flips to $t_{\alpha xy}$*; on α* consider the square 1*-2*-3*-4*.

Of vertex 3*, 3′ is obtained; 3*-5 is considered, the arc 5-6 of center $T_{\alpha z}$, 6-7 is halved in 7-8; 7-8 is rotated in 7-9, parallel to O-$T_{\alpha x}'$; 9-3′ is parallel to 5-3*, indeed, it is equal to it, since it is a horizontal of α, it is parallel to π = yz and therefore preserves its length.

Similarly, we can proceed to obtain the vertices 1′ and 2′, the 4′ is obtained by parallelism, since the opposite sides of the parallelogram projection of the square must be parallel to each other.

To obtain the 2′ you can use the line of the 3*-2*; it affects the $T_{\alpha z}$-$T_{\alpha x}$* and the $t_{\alpha xy}$* and these points are reported on the $T_{\alpha z}$-$T_{\alpha x}'$ and the $t_{\alpha xy}'$; the result is the line 3′-2′; the horizontal section that distances 2* from the $T_{\alpha z}$-$T_{\alpha x}$* remains unchanged and allows the 2′ to be obtained.

To obtain the vertex 1′, you can use the lines 3*-1* and the one joining the point 1* with the mean of 3*-2*; the corresponding intersects in 1′.

Consider for O the orthogonal (n_α) = O-10; the n_α' = O-11, on x′-z, is the direction of the projection of the axis of the octahedron orthogonal to the square section placed on α.

For the 12′, projection of the 12* center of the octahedron, the parallels to (n_α) and n_α' are conducted; on the first one, the 3*-12* semidiagonal is brought back to 12′-13, and on the second one, we get the 12′-14 projection by shortening according to the 13-14 parallel to 10-11.

With the 14-12′ = 12′-15, we obtain the last vertex of the octahedron projection; by joining the 14 and 15 with 1′, 2′, 3′, 4′, we finally have the oblique axonometric view of the octahedron.

Let it be reflected; The projections of vertices 14 and 15 on the x-y plane are obtained according to z on the parallel to x′ conducted for the 12′.

Now we begin to deal with the subject of oblique axial symmetry; for 10 we lead (d) = 10-16, φ being the angle 16-10-O between (d) and (n_α); d = 11-18 is the direction of obliquity with respect to nα, in projection and forms with n_α' the angle φ′ = 18-11-O.

Assume a horizontal α radius of 10-19; consider the 10-20 = 10-19 reduced to 11-21 on the $t_{\alpha xz}'$ according to the ($T_{\alpha x}$)-$T_{\alpha x}'$; to avoid overlaps harmful to understanding, translate the segments 10-19 and 11-21 into 23-22 and 23-24.

They are the conjugate radii of an ellipse 25, projection of a circumference of position α; the first is horizontal and the second has the direction of the line of maximum slope of α with respect to π = y-z.

On the plane, defined by n_α' and 24-23 parallel to z-x, consider for 24 the parallel to d′, coplanar and incident n_α' in 26; The projection of a right circular cone has been constructed, which has the base 25, the n_α' axis, the vertex 26.

Any generator of such a cone form with n_α' the angle φ; a′, incident α in P_o', let the projection of the axis of symmetry and be chosen parallel to 14-15, i.e., to n_α', orthogonal to α.

Obtain the symmetry of any point P′ of α, for example of the vertex 3′; consider the axial plane β_P' defined by a′ and P_o'-3′; the apparent angle φ_P' for 3′ depends on the direction of P_o'-3′.

Write P_o'-3′ in 23-(3); (3)-6 is the projection of the generator, whose direction, conducted for 3′, cuts the axis a′ into 27, forming the angle φ' projection of φ pertaining to Po′-3′.

The 3′-27 is reported in 27-$3_s'$; the $3_s'$, symmetrical of 3′, is a projection of the symmetry of vertex 3 of the octahedron; note that 3 belongs to α, while 3_s is in space.

Consider α the segment 3-12-1, which does not contain Po and is skewed with the axis a; the spatial symmetrical of the points of the segment does not form a segment, and the projection of the curve results in a curve; the midpoint 12 loses this quality.

The symmetrical figure of the flat square, in space, is the set of four humped lines; parallelism and proportion are lost; the intersection is preserved, at the finite and at the infinite.

Of the projection of the aforementioned square, only the diagonal 3-1 and the side 3-4 are schematically reported; already at this level, reading appears difficult.

It is suggested to complete the research; several questions are highlighted; what emerges from the determination of the symmetrical of points outside α, such as vertices 14 and 15? And, are the symmetrical faces at the eight equilateral triangles humped surfaces? Do problems of apparent contours come into play? What happens in the case of Po coinciding with the center 12 of the octahedron? And, for what positions, positions, directions, shapes, and sizes, can appreciable simplifications be obtained in the execution? And all outcomes on the pitch? What effects can be expected for relative movements?

Here, let it be reflected; a summary study is devoid of rationality, which runs aground within the limits of an oblique axial symmetry of a single point!

A few other considerations are useful; the axis may not be orthogonal to α; it may be parallel to α; the lines of the plane, parallel to the axis, give rise to symmetrical spatial lines, parallel to the axis; any other line of different direction exists, by oblique axial symmetry, a curve, helping to form a grid of lines and curves; it is a ruled surface; is it flat or convex?

Let us now propose, in oblique axonometry, with a view from below, defined by a triplet x, y′, z, O, and $u_y' = u$, a problem of orthogonal symmetry, with respect to a generic plane β.

Consider a generic plane α, on which a ring rests; it is a straight cylinder with a coaxial cylindrical cavity, bounded by two parallel bases, one of which is on α; the straight line is a circular crown, and the axial section is made up of two rectangles (Fig. 3.15).

The plane α is defined by $t_{\alpha xz}$, $t_{\alpha yz}'$, $t_{\alpha xy}'$; the traces on the axes are $T_{\alpha x}$, $T_{\alpha y}'$, $T_{\alpha z}$, to draw a circle on the plane α, it is necessary to determine a pair of its conjugate diameters.

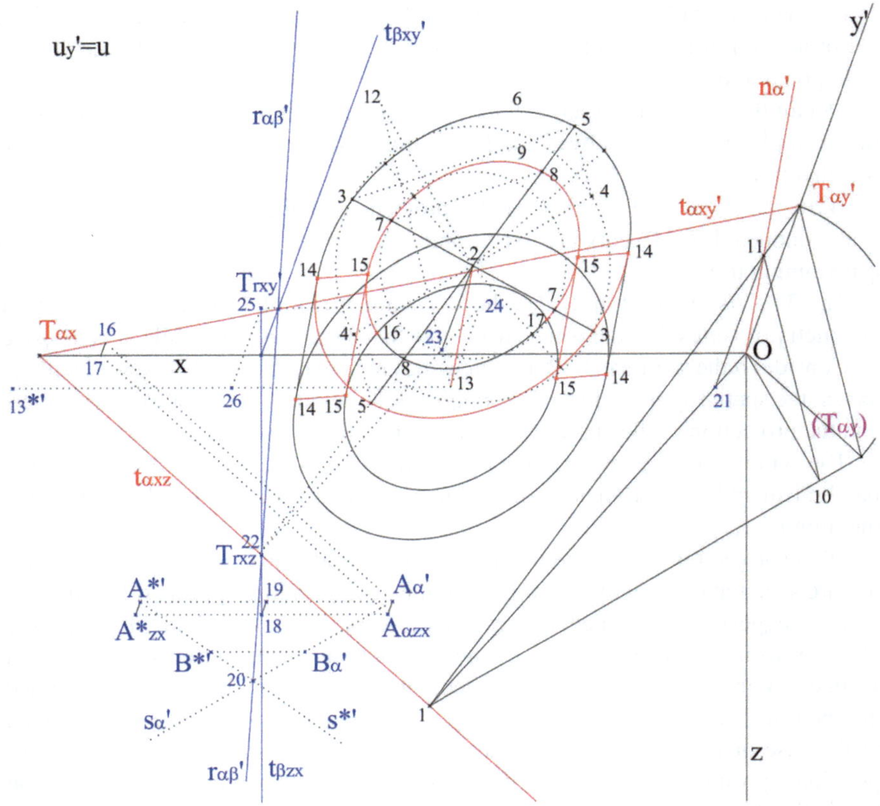

Fig. 3.15 Orthogonal symmetry, with respect to a generic plane β of a straight cylinder with a coaxial cylindrical cavity viewed from below, defined by a triplet x, y′, z, O, and uy′ = u

The projection of a diameter of α, parallel to $t_{\alpha xz}$, is in true magnitude; the conjugate, having the direction of the line of maximum slope of α with respect to π = xz, must be obtained in the direction and homological ratio.

Consider the track $T_{\alpha y}'$, which on xz projects into O; the O-1, orthogonal to the $t_{\alpha xz}$, is the projection on xz, in true size, of the $T_{\alpha y}'$-1; orthogonal to O-1 is O-($T_{\alpha y}$); 1-($T_{\alpha y}$) is the true greatness of 1-$T_{\alpha y}'$.

Once the first diameter has been chosen in 3-2-3, to obtain the second one, we consider the 2-4 parallel to the 1-($T_{\alpha y}$), and the parallel bars to the ($T_{\alpha y}$)-$T_{\alpha y}'$ for the points 4 defined on the parallel for the 2 to the 1-$T_{\alpha y}'$ the diameter 5-2-5 conjugated to the 3-2-3.

The axes are obtained and ellipse 6 is drawn; Once the diameter 7-2-7 has been established, the conjugate 8-2-8 and the ellipse 9 are obtained by homothety; The projection of the second base of the ring is obtained by translation of the aforementioned ellipses according to the direction of the n_α projection.

Consider the O-10 orthogonal for O to the 1-($T_{\alpha y}$); the parallel to the ($T_{\alpha y}$)-$T_{\alpha y}'$ for the 10 identifies on the 1-$T_{\alpha y}'$ the point 11, projection of the intersection on α of the n_α'; take into account that O-10 has been reduced to O-11.

Establish, in true magnitude and parallel to 10-O, the height 2-12 with respect to α, between the base on α, already drawn, and the other, with the parallel one; on the parallel to the n_α', conducted for the 2, the extreme 13 is identified by means of the parallel to the 10-11; the 2-13 is the projection of the height.

The 13 is the center of the second base, which is obtained by translation according to the n_α'; for the extremes of the diameter conjugated to the direction 2-13, the projections of the apparent contour generators of the two coaxial cylindrical surfaces of the ring are conducted.

Of the above-mentioned generators, only the outer ones 14-14 are visible; the inner ones 15-15 are not visible; the upper base, which is the lowest in space, is all in sight; of the base belonging to α, the outer ellipse arc 14-5-3-14 and the inner one 16-8-17 are visible.

Once the construction of the ring is complete, consider a plane β of orthogonal symmetry; the traces $t_{\beta xy}'$ and $t_{\beta zx}$ define a plane β parallel to yz, which intersects α in $r_{\alpha\beta}'$ of traces T_{rxy}', T_{rxz}; this line of α, belonging to β, is joined to its symmetry.

From the point A_α', we derive $A_{\alpha zx}$, through A_α'-16, parallel to $t_{\alpha xz}$, 16-17 parallel to y′, 17- $A_{\alpha zx}$, parallel to $t_{\alpha xz}$, A_α'-$A_{\alpha zx}$ being parallel to y′.

The direction of the orthogonal symmetry to β is given by x; for $A_{\alpha zx}$ the parallel to x, which intersects β in 18, is conducted on xz; the symmetric is again on xz in A^*_{zx}; of $A_{\alpha zx}$-A_α', parallel to zy and a β, the symmetric is A^*_{zx}-$A^{*\prime}$, again parallel to y′.

Consider a line s_α' for A_α'; its point 20, common to $r_{\alpha\beta}'$ is united; $s^{*\prime}$ is given by 20-$A^{*\prime}$; a point B_α' of s_α' has as its symmetry the point $B^{*\prime}$ on $s^{*\prime}$.

The spatial points of α have the spatial points of $\alpha^{*\prime}$ as symmetric; they are linked by an oblique affine spatial homology, having $r_{\alpha\beta}'$ as its axis and the direction of the x-axis as its center.

The image of the symmetry of 6 is obtained by points, as for A, that is, by the conjugate diameters 5*-2*-5* and 3*-2*-3*, on α^*, symmetric of the conjugate diameters 5-2-5 and 3-2-3, of α; the relative completion is left to the scholar.

Reflect on the explicit and implicit "whys" of the construction, on the possible complications resulting from the different interrelationship of the initial data or fixed in the process. The learning of the subsequent fluent passages to the results must be the result of skillful weaving of connections, references, comparisons with previous knowledge that supports or ratifies the thought and thinkable of the "if" of the emerging integrations.

The projection on xz of point 11 is 21; conducted for the 2 on α the parallel at 11-1 for the 2 identifies the 22 on the $t_{\alpha xz}$; the parallel for the 22 at 1-21 meets in 23 the parallel for the 2 at 11-21, and in 24 the parallel for the 13 at 11-21.

The parallel to x for the 24 of xz obtains the 25 on the $t_{\beta zx}$; the parallel for the 13 to the 24-25 identifies the 26 on the parallel to the 24-13; the 13-26 is returned to 26-13*′; the symmetrical ellipses of the base on α^* already determined are translated with the center to 13*′.

The completion of the figure is left to the scholar; the ellipses and tangents have been traced; the visible parts are highlighted; the resulting image of the last ellipses can be obtained by homology, after determining the new axis, intersection of β no longer with α but with the parallel plane considered for 13.

It is an extremely useful experience to imagine the results to be obtained or completed and to verify the correct prediction, that is, to be persuaded of the reasons that could have given rise to them.

It does not seem unjustified to insist on the utmost attention and on the spirited curiosity of eminently formative certainty, in preparing to give an answer to a proposal to be solved as a problem and not as a problem or exercise.

The symmetry with respect to a plane can be non-orthogonal but oblique; also, in this case, a homological aid is possible for the figures belonging to a plane.

The choice of the data of the reference system of the oblique axonometric projection, the position of the planes involved, and the direction of symmetry must also be weighed to avoid the desired outcomes do not fall out of bounds.

It should be noted that the concept of symmetry with respect to the plane can be mistakenly confused with the phenomenon of reflection on specular plane surfaces; in the future treatment appropriate to reflected images, the inescapable diversity of settings, purposes, and independence from imaginable promiscuity can be highlighted.

Chapter 4
Volumes in Oblique Axonometry

Abstract This chapter extends oblique axonometric techniques to the representation of complex three-dimensional solids. Starting with polyhedra and progressing to ruled and curved surfaces—such as helicoids, quadrics, and toroids—it presents methods for accurate geometric construction and projection. The chapter emphasizes the importance of visual reasoning and proportional relationships in creating readable, accurate, and scalable images. It also addresses aggregate forms and symmetry operations. Special attention is given to the implications of projection direction on the deformation and readability of the volumes. This section consolidates and applies the concepts introduced in earlier chapters to volumetric geometry.

Polyhedra

In infinite geometric space, it is possible to isolate a portion of it; this space is defined if it is delimited by an envelope; it consists of flat surfaces or humps, and the latter can be concave or convex.

The envelope separates the portion considered from the rest of the whole; one or the other or both can be incommensurable, infinite, indefinite, that is, the infinite portion and the finite other, or the finite portion and the infinite remainder, that is, both finite, being a finite part of another finite space, part of an infinite remainder.

Think of an endless tube, with a straight, constant, polygonal section: the envelope, with flat faces, can be understood as a delimitation between two infinite spaces, one full, the other empty.

If the full one is the pipe, the delimitation made by the envelope is external; if the full one is the outside and the pipe is empty, the delimitation is inside.

For the other hypotheses, the scholar should try to reconstruct corresponding examples; in the end, you think, to simplify, of a closed, finite, full volume, contiguous to the rest, infinite.

It appears to be delimited by an envelope, which can be made up of a homogeneous surface, such as the sphere, the ellipsoid, not the paraboloid or the hyperboloid, or of non-homogeneous, composite surfaces, intersecting according to single or multiple

L. Inzerillo, *The New Horizons of Oblique Axonometry*, UNIPA Springer Series,
https://doi.org/10.1007/978-3-032-07475-1_4

curved lines, in turn homogeneous and non-homogeneous, or intersecting at vertex points or nodes.

Let it be reflected; the aforementioned intersection curves can be virtual, in particular situations of tangency; this is also evident for nodes; think also of the cases in which a vertex or node is improper, for which the space enclosed in the envelope is infinite, and the envelope appears open to infinity.

The envelope can consist of all flat surfaces, in a minimum number of four, called faces; These are polygons with a convex or concave perimeter, which share straight segments called sides, which in turn incident at points called vertices.

Nothing prevents us from thinking of mixed envelopes, consisting partly of flat surfaces and partly of curved surfaces, such as a prismatic base surmounted by a pavilion vault with cylindrical spindles.

The scholar receives a suggestion; take the opportunity to reflect, understand, and propose many other examples; it is not the example that leads to the concept; the latter is the fruit of thought, the other is the secondary and unnecessary indication of the conquest, of the consolidated possession of the concept, and is not its validation.

The analysis of the formation of a flat-faced envelope should start by observing a vertex and the polygons that share its belonging; The vertex encloses a solid angle.

The sum of the plane angles of the faces competing at the same vertex must be less than 360°; the number of faces competing at the vertex has only the minimum limit of three; the minimum number of solid corners constituting the closed envelope is four; the envelope entirely of flat faces just described is called a polyhedron.

The infinite number of possible polyhedra can be divided into groupings; in each one, exclusive characteristics and geometric properties are highlighted, gradually reported, or premises.

Two planes always intersect in a straight line; if they are parallel, the line of intersection is improper; three planes intersect according to three lines; they have in common a point, proper or improper, the vertex of a solid angle.

A fourth floor, which does not contain the aforementioned point, intersects the trio of floors and isolates a multifaceted volume with four faces and four vertices, each opposed to one face; These are triangular faces.

This polyhedron can have triangular faces that are all different, or two or three or all four are the same as in the tetrahedron; The analysis must also be extended to the edges and all significant angles, internal and external.

Three parallel edges, two by two, forming flat faces, form an infinite prismatic surface, whose straight section is a triangle; a plane dissects and limits the still infinite surface.

A second floor sections the aforementioned polyhedron, whose volume is finished; it is possible to make the two planes parallel, to be right with respect to the edges; a polyhedron was formed with three rectangular faces and two triangular faces.

If the rectangular faces are equal, the triangular faces are equilateral triangles; these, if the rectangular ones are orthogonal to the support surface, are commonly designated as the bases of the lower and upper volume; they are not if the support is on one of the rectangular faces.

There is no limit to the number of rectangular faces; the corresponding bases are polygons having the same number of sides. If the rectangular faces are equal to each other, the base polygon is a regular polygon.

It is not enough; the basic polygon can be with n sides, equal or different, convex or concave, intertwined, with multiple vertices, with cyclic development, etc.; the basic perimeter can be understood as a guideline.

A generating line, or rather, a generating segment, moves continuously, maintaining its own extremity in the aforementioned direction; the polyhedron in question is born.

The faces with infinite rectangular bands thus generated form a striped surface. Note that the polyhedron is still such if the bases are not parallel or if the faces are parallelograms; These are oblique prismatic polyhedra.

The typological analysis of polyhedra cannot be exhaustive; Imagination can supply and increase without end the personal knowledge of polyhedra, of which natural and constructed reality is improbably and inevitably very rich.

Therefore, a conceptual investigation is remarkably formative, above and outside of an exemplary curiosity engaged in non-conformist hypotheses.

The vertices of any right-sided polygon, or of a directrix, flat or not, of straight segments, can be joined to a point called the vertex, defining a space, finite or open to infinity, a polyhedron called a pyramid.

Generation can be understood to be realized by the generating line passing invariably through the vertex and, which runs on each of the infinite points of the line with rototranslational motion; an ordered sequence of ruled surface distinct from that examined for the prismatic polyhedron.

In addition, it should be observed that in the prism the generator invariably contains an improper vertex point, while in the pyramid the vertex is proper; in the pyramidal case, two distinct pitches are generated, opposite the vertex.

Any limitation with two sectioning planes defines a double pyramid, unless one of them contains the vertex; if the two planes are on the same side of the vertex, a truncation of a pyramid is obtained.

It should be noted that the guideline can be opened, both in the case of the prism and the pyramid; even the base parting off may not constitute a base flat face.

Think of all, or almost all, the possible truncations forming a non-flat spatial polygonal, consisting of rectilinear sections or coplanar arches, with a rectangular face in the prism and a triangular face in the pyramid, free, however irregular, even star-shaped.

Also interesting and necessary is a spontaneous incursion into the field of associations, budding, combinations, compositions, and aggregations between polyhedra.

Nor can particular attention be paid to the intersections, interpenetrations, chamfers, fillets, drillings, external and internal hollows between polyhedra, as well as the predictability of the consequences of any modeling or modifying intervention.

In essence, the scholar must not submit to the current educational simplifications, which see the geometric knowledge on polyhedra treated, or rather, mistreated.

In them we range, with sufficiency and presumptuous satisfaction, from the cube to the regular straight hexagonal pyramid, even proposing its development, and to exceed, an elementary example, sensationally original, of a flat section of recommended position and intersection with a recklessly particular, horizontal straight line of direction; the scholar, the arbiter of his own knowledge, should put such proposals on the index.

On the reference x′, y, z, O, with $u_x{}' = u$, (Fig. 4.1), we want to draw an antiprism with eight isosceles triangular faces; the two square bases are rotated by 45°; the first belongs to a plane α, vertical by z, and of a $t_{\alpha xy}{}'$ trace.

The vertical diagonal A′-B′-C′ is in true size; the other, D′-B′-E′, parallel to the $t_{\alpha xy}{}'$, is obtained by similitude; if a point P′ of coordinates P′-1, O-1 is chosen on a $t_{\alpha xy}{}'$, O-2 is the true length of O-P′, which is reported in O-e; A′-D′ is parallel to P′-3.

Consider 4-5, 6-7 = 4-5 and 7-8 orthogonal to O-7 in sequence; let the distance between the two bases be 7-10; let 12-13 = 11-10; 9-4-13 is the direction of height B′-F′; for the 13th, the $t_{\beta xy}{}'$ parallel to the $t_{\alpha xy}{}'$ is conducted; $t_{\beta yz}$ follows.

For F′ we conduct the parallel to $t_{\beta xy}{}'$ and the vertical; on them, by intersection with the parallels to B′-F′ conducted for the four vertices A′, D′, C′, E′, the points 14, 16, 15, 17, mean of the four sides of the base, are identified; the vertices are G′, H′, I′, L′.

The second base appears translated, rotated by 45°, as well as enlarged; by joining the vertices of the two bases, a polyhedron is obtained; if the two bases had retained their size, an antiprism with a square base would have been obtained; instead, it is a deltahedron.

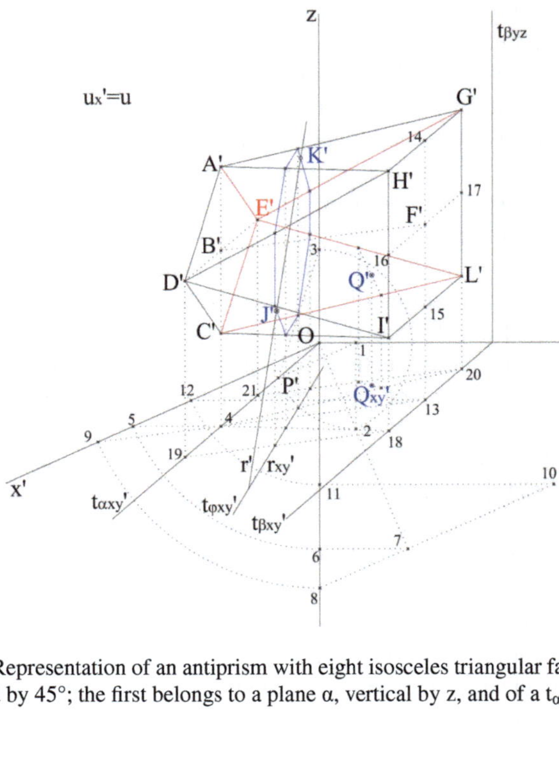

Fig. 4.1 Representation of an antiprism with eight isosceles triangular faces; the two square bases are rotated by 45°; the first belongs to a plane α, vertical by z, and of a $t_{\alpha xy}{}'$ trace

The faces in view are the triangles A′-H′-G′, A′-H′-D′, H′-D′-I′, I′-D′-C′, I′-H′-G′-L′; the rest is hidden; the projections on xy of the edges are, respectively: 18-19 of H′-D′ and D′-I′, 18-4 of A′-H′ and C′-I′, 4-20 of A′-G′ and C′-L′, 20-21 of E′-G′ and E′-L′; those of the two bases are 19-4-21 and 18-13-20.

We want to know if Q belongs to the face C-E-L; any secant for Q′ intersects the sides E′-L′ and C′-L′; the verticals for these points intersect the relative projections 21-20 and 20-4; the joining of these points contains Q_{xy}'.

The test can be performed starting from Q_{xy}'; if Q′ does not belong to the joining points of the relative secant, then Q is not on the face; be careful; in this case, if you request to raise Q′, it is external to the polyhedron, and vice versa.

If the search is made for the face A-H-D, and Q′ must go up, it is internal; it is necessary to know how to analyze the envelope; you have to have a sense of space; The eye looks, materializes the image, hypothesizes movements, decides relative positions, and knows how to move the elements to obtain relative positions.

Given the projections r′, r_{xy}' of a generic line r, we want to determine the intersection of r with the polyhedron; consider the plane φ, vertical for r; the verticals for the points of intersection of r_{xy}' with the projections on xy of the edges, already obtained, identify on the relative edge the vertices of the section of the polyhedron with φ, in this case, octagonal.

The r, belonging to φ, in this case, intersects the said section at two points, J′ and K′, crossing the polyhedron through the segment J-K; the first belongs to the face C-I-D and the second to A-G-E.

The intersection between a polyhedron and a generic plane offers the opportunity for an exceptional commitment in building the image of the polyhedron, following a thought; a form is expressed, and its properties are expressed with deep reflections.

To give an idea of the problem, just think of a straight regular hexagonal pyramid with the base resting on the horizontal plane xy, being $u_z' = u$, a bold example reported only in the best texts, marking the border with the unknown.

This exercise, even if used to directly derive a generic plane section from it, by means of homology, in 45° axonometric projection, with respect to the one proposed below, (Fig. 4.2), is an exercise for beginners.

Let a hexagon of vertices 1, 2, 3, 4, 5, 6 be assigned on xy, with center in V_{xy}; at the vertex V′ six edges converge, like 1-V′; the diagonals of the hexagon represent its projections on xy, such as 1-V_{xy}.

But 1-V′-2 is not the face of the pyramid; consider on the bisector V_{xy}-3 the 4, projection of 5, which exceeds the V′-3; the V′-5-6 is replaced by the broken V′-5-3; six triangular faces are highlighted as 1-5-2 of 1-4-2 projection.

They would be cut out as part of the faces of a different coaxial pyramid, having the same base and vertex on the axis at a higher altitude than that of V′; this vertex is the intersection of the 3-5 with the V_{xy}-V′ axis, out of bounds; the 5-V′ edge projection is 4-V_{xy}.

The two faces 1-V′-5-1 and 2-V′-5-2 are triangular, projected 1-4-V_{xy}-1 and 2-4-V_{xy}-2; let it be reflected; triangles are part of triangles 1-V′-6-1 and 2-V′-6-2; edges such as 1-V′ are not ridges of a convex solid angle, but bottoms of a concave angle.

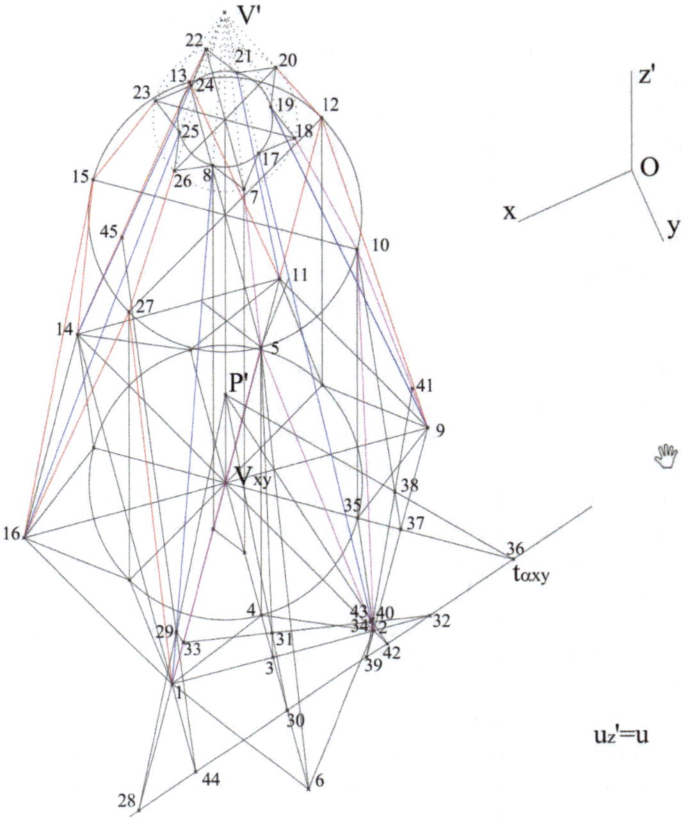

Fig. 4.2 Section of hexagonal pyramid: Truncation by a generic plane with visible face results

Of course, one perceives how much those who rely on appearance can dismiss, if they are conditioned by habits determining inadmissible preconceptions, wanting to declare themselves experts in the geometry of drawing.

The xy projection of 1-V′-6-1 is 1-V_{xy}-6-1; chosen the 7 on the 5-V′, the horizontal 7-8 parallel to the 6-1 is the side of a six-pointed star, horizontal flat section at the altitude of 7; it appears rotated.

Reading the polyhedron requires attention; 8-7-5-1-8 is in sight and vertex 5 is prominent; the 17-7-5-2-17 is in sight; the 17-18-10-2-17 is also in sight; the 19-18-10-9-19 is hidden; the 19-20-12-9-19 is partially visible and the part 9-10-18-19-9 is hidden; the faces 21-20-12-11-21 and 21-22-13-11-21 are hidden; 24-22-13-11-24 and 24-23-15-14-24 are hidden; the 25-23-15-16-25 and the 25-26-27-16-25 are in sight; the 8-26 27-1-8 and the 8-7 5-1-8 are also in sight.

The interest of being able to follow with the numerical reading and the relative increments of each subsequent segment requires the preservation of each graphic increment, but this is to the detriment of the emergence of the lines of the envelope;

it is not advisable to punctuate either the construction lines or those not in sight. The drawing should be greatly enlarged.

We want to complicate the drawing, considering a flat section of trace taxy and point P′ on the V_{xy}-V′ axis; we proceed by oblique affine homology, whose axis is the $t_{\alpha xy}$ and whose improper center is given by the direction of the axis connecting the pair of homological points P′ and V_{xy}, projection of P on xy.

We can start with any element; consider edge 8-1 and its counterpart on xy V_{xy}-1; a V_{xy}-1-28 corresponds to 28-P′, which identifies 29; to obtain the section on the 1-5-2 face, V_{xy}-4-3-30 is considered; the corresponding 3-5 identifies 31 on 30-P′; 1-3-2-32 corresponds to 32-34-31-33; the result is also 29-33.

We continue with V_{xy}-35-337-36, with 37-10, which finds 38 on 36-P′; the 9-37-2-39 corresponds to the 39-38 which intersects the 2-10 and the 9-10 respectively in 40 and 41; the V_{xy}-2-42 corresponds to the 42-P′ which obtains the 43 on the V′-2; it must be combined with 34 and 40; The 29-33-34-43-40-41 traverse is a part of the section line.

To obtain the rest, one can proceed according to the ways illustrated; another route is indicated by 14-1-44 and 44-29-45; 14-V′ is opposed to 2-V′, so 43-P′ identifies 14-V′ as 45. All the homological routes, applied with great care, simplify the determination of the section line and facilitate the control of its correctness.

The projection on xy of the upper truncation is not reported, just as the section is not completed with a plane α, and the entire line is not transformed into a dotted line, both of the projection of the polyhedron on xy and of the parts not visible.

The guiding concept of the treatment is and must remain the training in know-how and not the image product deriving from the data; Only in this way can the certainty of resolution and the freedom of proposal without nightmares be conquered.

All this is reiterated, proposing a different envelope (Fig. 4.3), which, and it is always to be expected, inevitably engages reflection, reasoning, attention, and continues towards the goal, drawing on the sources of knowledge consolidated thanks to previous experiences.

Adopt in a top view, a reference x, y′, z, O with $u_y{'} = u$; place on xz = π the pentagon of base 1-2-3-4-5 of a prism, considering the part 3-4-5-6-7, which emerges from the xy plane; points 8, 9, 10, let the projections on x of vertices 7, 4, 3.

Think of a right prism, having the aforementioned base on xz; The projection of its edges on xy are parallel to y′ and their length is in true magnitude equal to 9-13.

The second base is vertical and parallel to xz; in projection on xy it is defined by 11-12-13-14-15 and in projection on xz by 16-17-18-19-20; The horizontal edges, projected on xz are 6-20, 5-19, 4-18, 3-17, and 10-16.

Think of abandoning the banality and keeping the true length 8-14 of the edge, want to move the base on xz parallel to x so that the relative projection on xz overlaps the 20-19-18-17-16.

Stop; let it be reflected; the 5-19 becomes only the 19; in projection on xy it cannot be 21-14, since the edge 8-14 in its new position, which seems orthogonal to x, stands out from 21 but, to remain equal to 8-14 = O-22, it will appear shortened and equal to O-23.

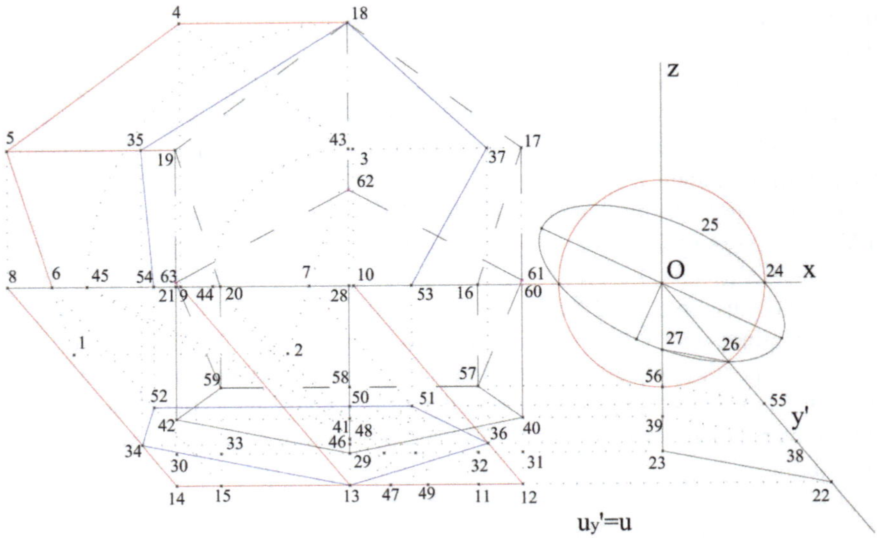

Fig. 4.3 Truncated prism projection: Deformation and section reading under oblique axonometric projection

To obtain O-23, we must consider the projection of a circle of xy, for example, of center O and radius O-24; by homology, we obtain the ellipse 25; the radius O-26 according to y′ becomes O-27; by similarity according to 26-27, the length O-22 = 8-14 appears shortened and equal to O-27.

Therefore, the 9-13 becomes 28-29; in projection on xy, the base is given by the segment 33-32-31-29-30-33 parallel to x; the prism has two equal and parallel bases; The edges of its lateral surface, projecting on XY, all appear the same.

Composing the two projections on xz and xy in space, the result is 17-61 = 60-40, 18-62 = 28-29, 19-63 = 21-42; 62-61 and 62-63 are not parallel to 17-18 and 18-19; 44-59 and 16-57 remain stationary.

The oblique base 62-63-59-57-61 62 and the 18-17-61-62-18 and 18-19-63-62-18 are in sight; The two side ones 17-61-57-16-17 and 19-63-59-20-19 and the base on XZ are hidden.

The sensation aroused by the appearance is that the resulting prism is right, and instead it is oblique; visual perception, and not rational perception, suggests as an oblique prism what is in fact right. The composition of the two projections in space relating to the right prism, which has edges parallel to Y′, has not been carried out.

The real orthogonal distance between the planes of the bases of the right prism is 8-14, but to the unwary, it appears 21-14, because it does not take into account the obliquity of y; in the oblique prism, the aforementioned distance is actually decreased.

To read and understand the real spatial model in the two cases, it is necessary to put the appearance aside; Necessarily, one must reason, even if it is hard!

We want the surprises to have no end; it is proposed, without prejudice to the base 20-19-18-17-16 on xy, to conduct a plane α for the vertex, of projections 29, 18, inclined with respect to xz, so as to obtain the track on xy parallel to x.

Establish the 13-34 and 18-35 on the left and the 13-36 and 17-37 on the right in the original prism; from 34-36-38, we arrive at 38-39 and 39-40-41-42; 34-46-36-38, 38-39, 39-40-41-42, 34-46-36-38 are considered.

By similarity, we obtain 50 with 46-47 and 43-44-47-48 and 18-45-48-49-50; and from 50, we proceed to 52-50-51-55-56-57-58-59; The projection on XY of the second base of the oblique prism is given by 42-29-40-57-59-42.

From 41, we can also reach 58 by traced but not numbered similarity; The inclination of the base does not change its projection on XZ, which remains the 20-19-18-17-16-20: therefore, the model changes, and a projection may not reveal it.

The modification of the model on site, operating on the projections, constitutes a rich source of reflections, of countless "ifs" and "buts"; concentration of thought is an effective guarantee of reliability and operational control.

The image cannot answer the naïve question of what it represents, if it is not read and understood; After the acquisition of the object, the image becomes communicative and exclusive, in the sense that it becomes capable of revealing the numerous messages it conveys to the user.

It is an image built in order to stimulate and provide a further, high, qualifying and qualified training of the expert in the geometry of drawing, and not of the expert designer.

Think, for a relative comparison, of those who are able to design new functions of an IT tool and those who learn to use it with the competence of a user.

Such papers are and must remain open to the Scholar, the Teacher, the Candidate for the Teaching of Drawing; they do not serve to strengthen the practitioner of computer image production; the latter would be confused, dissuaded, burned.

With reference to x′, y, z, O, with $u_x' = u$, we want to illustrate an example of intersections between two polyhedra, for example, between a truncated pyramid (Fig. 4.4) and a prismatic cavity.

The pyramid is intended to be oblique with a square base resting on xz; the projection of the base must be constructed, after reversing xz on yz $= \pi$; Given the inverted base of vertices 1-2-3-4, the aforementioned projection can be drawn from it in different ways.

Since O-5 = O-6, 5-6 is the direction of the center of the z-axis homology; considered 4-1-7 and 5-1-2-8, on 6-8, you get the vertex 10 by leading the parallel to 5-6 by 1; it is repeated for the other summits.

The 10 can be obtained by means of the 1-9, 9-10, 1-10; it seems a different way, but it is still the homology already used; in determining the additional vertices 12, 15, 14, symmetry with respect to the center 16, parallelism, similarity, etc., can be applied.

Define the vertex 17 of the pyramid, choosing projection 16-17 as the oblique axis and 20-19 as the projection, on xy, of the relative projection; The projections on x′ of vertices 14, 10, 12, 15 are 21, 18, 23, 22.

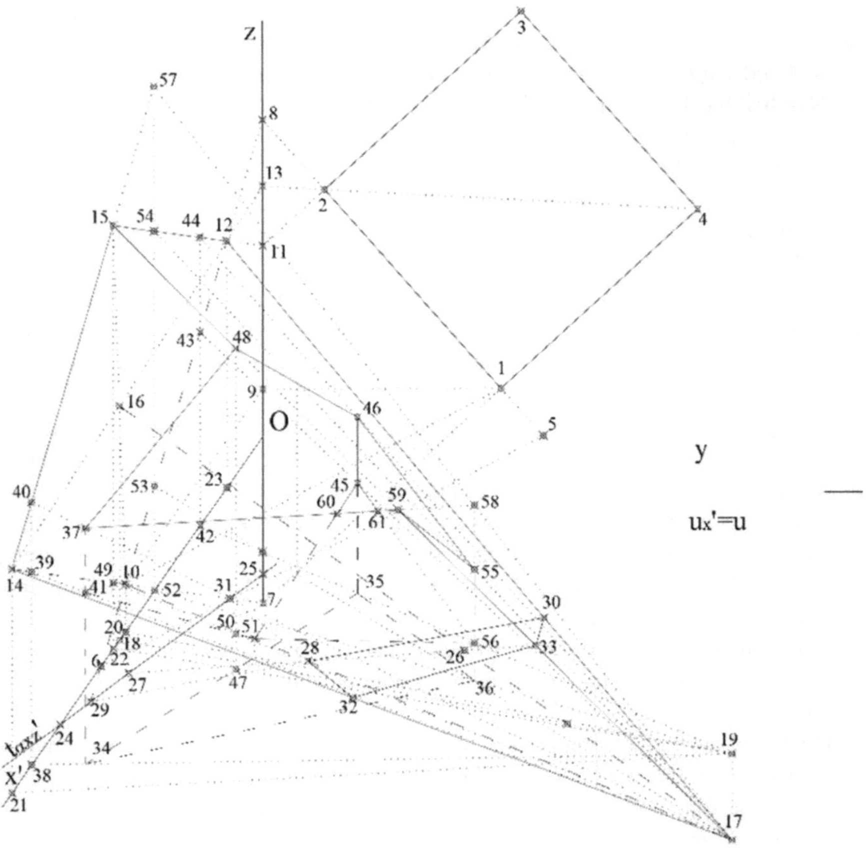

Fig. 4.4 Intersections between a truncated pyramid and a prismatic cavity

The top of the base should join with the top 17; the projection of the polyhedron is obtained; the faces 12-17-15 and the face 15-17 14 are clearly visible; the rest, reported with a dashed line, is hidden; By joining 19 with 21, 18, 23, 22, we obtain the projection of the projection of the pyramid on the XY plane.

Let us want a flat section of the pyramid; the plane α of section is defined by the trace $t_{\alpha xz}' =$ 24-25, and by point 26 on axis 16-17; the vertices 28 and 33 of the section on edge 10-17 and 15-17 are obtained by means of 15-16-10-27 and 27-26; 14-10-31 and 31-28 obtain 32; the fourth vertex, 30, it is obtained on 32-26.

Note that the pyramid is intersected by xy; this section has not been determined; If we do not take into account the part of the pyramid with vertex 17 and base 28-32-33-30, we have a truncated oblique pyramid with a square base on xz.

Proceed to extract the said volume; intersect it with the empty right prism, having the triangular base 34-35-36 on xy; The vertical plane, having track 19-34-38 on XY, and vertical track 38-39-40 on XZ, intersects the pyramid according to generators

40-17 and 39-17, identifying on the vertical edge for 34 the points of intersection 37 on face 14-17-15 and 41 on face 14-17-10.

The vertical plane, having track 19-35-42 on xy and vertical track 42-43-44 on xz, intersects the pyramid according to generators 43-17 and 44-17, identifying on the vertical edge for 35 the points of intersection 45 on face 10-17-12 and 46 on face 15-17-12.

The vertical plane having track 34-35 on xy intersects 22-19, projection of 15-17, in 47; the relative vertical identifies 48 on 15-17; therefore, be very careful; you cannot combine 37 with 46; The outline of the section is 37-48-46.

The 22-15 intersects 14-10 in 49 and the 49-17 generator intersects the 47-48 in 50; the 41-50 reaches the 10-17 in 51, which joins with the 45; therefore, the face of the prismatic cavity under examination, common to the pyramid is 37-48-46-45-51-41-37; The sides of the section are all on the faces of the pyramid except 41-37 and 45-46, which run through the trunk.

Similarly, proceed for the vertical edge, which has track 36 on xy; consider the vertical plane that has track 19-36-52 on xy and 52-53-54 on xz; on generators 54-17 and 53-17, the vertical edge obtains 55 and 56; 55 can merge with 46 but not 37.

Track plane 52-54 intersects the plane of face 14-17-15 off-face along virtual generator 57-17, which defines 58 on the vertical edge for 36; 58-37 gets 59 on 15-17; The edge of the upper hole is 37-48-46 55-59-37.

The 56-51 is on the face 10-17-12; of the 45-51, only the 45-60 section is in sight, and of the 45-56, only the 45-61; the 45-60-61-45 is a small part of the through-hole due to the hollow prism; with a lot of patience and attention, the proposed research was finally concluded.

In the field of polyhedral cavities, there can be placed not only internal, generic, coaxial, concentric, homothetic, dual, etc., but also those open to the outside and those passing through, as in the case illustrated above.

These can be simple or multiple cavities; Combinations can occur in aggregates with central groups, in interpenetrated sequences, structured in cells or modules, joined, chamfered, etc.

Let your imagination try to imagine virtual cavities; the frame protruding from a wall has a straight-shaped section; from the rough frame having a given straight section, it is possible to arrive at the final shaped one by removal, by successive cavities, etc.

The generation of protrusions can be interpreted as a function of a machining process, which does not add the overhanging volume but removes a volume for which the substance excluded from removal results in a full cavity.

Unimaginable fields of research are conceivable; there are considerations of use, not geometric, of so-called disposable solid cavities, or of moldings with over thicknesses, of inlays of mixed marble floors, of figurative mosaics, etc.

If the suggestions open new windows to peer into the unknown, into the unconscious, with curiosity, with suggestion, with the increase of one's own potential of knowledge, more and more, the scholar should not give weight to the example but collect the methods, the concepts, and meditate on the opportunities that knowing how to solve an example described by others does not raise him much, reflect on the function of exercise as an optimal opportunity, as an incentive.

We want to represent (Fig. 4.5) a large road work, a viaduct on straight prismatic pylons with a rectangular section of 10 m in the longitudinal direction according to the axis of the viaduct and 5 m in the transverse direction orthogonal to the aforementioned axis; The center distance of the pylons for the center of the section of the shutter on XY is 40 m.

Let us represent no less than two spans on a straight axis and two successive straight spans, each of which has the axis forming the angle with the section preceding it, such that the four consecutive sections constitute the sides of a regular polygon of 20 sides.

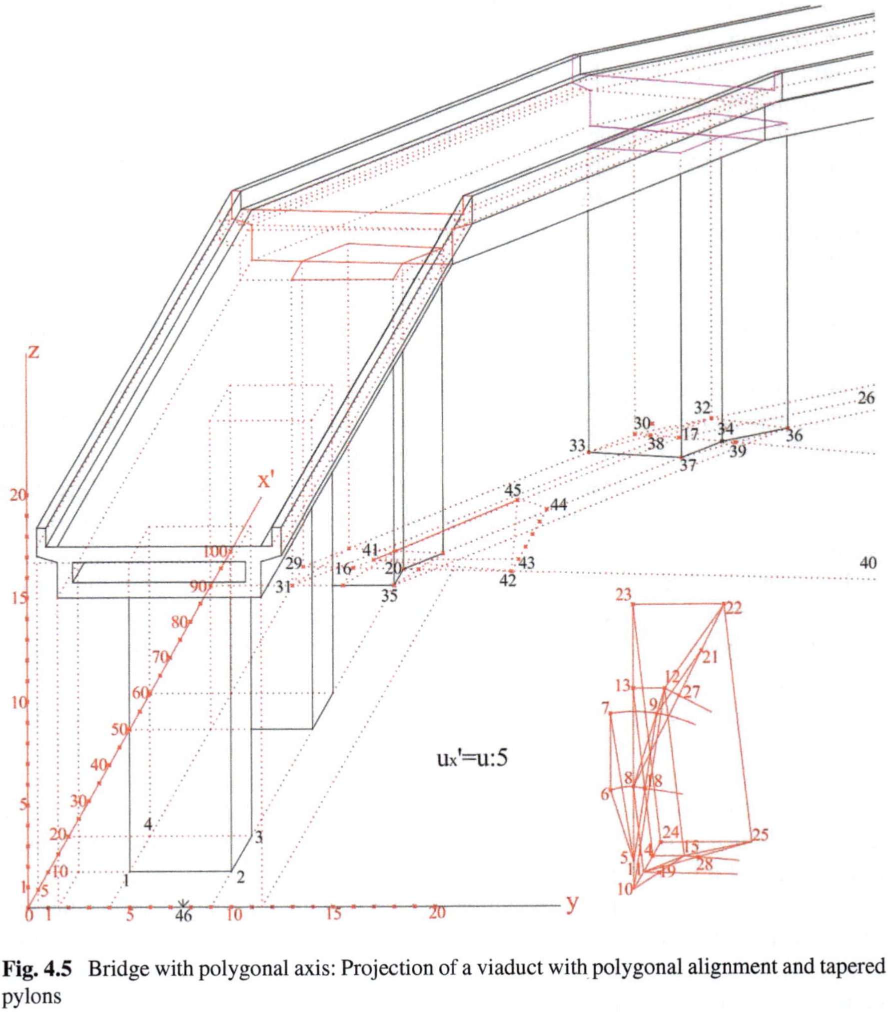

Fig. 4.5 Bridge with polygonal axis: Projection of a viaduct with polygonal alignment and tapered pylons

The height of the pylons is 15 m and the caisson above is with a hollow rectangular prismatic section of a constant thickness of 75 cm., width 10 m. and height 250 cm. with tapered section sidewalks, having the cantilever from the caisson of 1 m. and the height of 50 cm at the interlocking and 30 cm at the free end, including an adequate parapet.

Furthermore, if the value of the graphic scale on the reference y-z frame is around 1:200, in order to have a sufficient depth of field to avoid foreseeable inconsistencies off-screen, $u_x' = u:5$ is adopted.

Choosing the unit u on y and z in 0-1, the unit equal to 1:5u is set on x′, so that on x′ it is $5u_x' = u$ and the graduation is set on x′; this reaches the value at depth equal to $200u_x'$; The base of the first pylon is drawn in 1-2-3-4, being 1-2 = 5u parallel to y and 1-4 = $5u_x' = 10u$.

Note that 1-2 appears longer than 2-3; in reality, it is not so; Those who rely on appearances forget to observe and reason about the image, to use the eye to reflect and extrapolate from what they see what, in fact is.

The head of the caisson with cavities and overhangs appears strangely off-center; is the effect of the advancement of the starting section; The indications of the longitudinal and transverse slopes of the road surface, the gutters, the manholes, the screen railings and curb stones, the various pipes, pipes, horizontal indications, etc., have been neglected, but are very important.

The road axis, after the two straight spans, has a detour; the second and third are consecutive sides of a regular polygon with 20 sides; The construction of the polygon is set up, drawing two consecutive sides 5-6 and 6-7 on the xy plane, overturned on yz.

Rigidly report, rotating around 5 the 5-6 in 5-8 and the 5-9, being 8-9 = 6-7, with 5-8 parallel to (x) = z; let 8-10 be equal to one-eighth of the distance between the pylons on y(x) overturned by xy; the 10-8 is projected on the direction x′ conducted for the 10 in 10-13.

On the 8-9 direction, let 8-12 = 10-8; the direction 10-8 is that of (x) and the direction of 10-11 is that of x′; according to Y, the 12 moved away by 12-13 from the 10-8; the 13-14 parallel to the 8-11 and the 14-15 equal and parallel to the 13-12 identify the 15 of x′y.

In projection, 10-11 and 11-15 represent the directions of one-eighth of the two consecutive sides of the 20-sided polygon; Moving vertex 11 to 16, the center of the pylon, and eightfold 11-15 to 16-17, places the center 17 of the next pylon.

The 8-18 is the bisector of the angle between the two consecutive sides in overturned; the corresponding 11-19 is the projection on x′y; bring it back to 16, obtaining 20.

Reproduce the inverted triangle 10-8-12 in 8-12-22; the projection 11-15-25 is derived, which is reported, eightfold, in 17-26; from 12-27 we get direction 15-28, which is returned to 17; on this, the parallel for the 29th to the 16-17 identifies the 30th.

For the 30th we consider the parallel to the 17-26 on which, the parallel for the 31st to the 29-30, obtains the 32; the parallel bar for the 32 to the 11-28-25 gets the

33 on the 29-30; on 30-17 the parallel bar for 20 to 29-30 gets 34; on the parallel bar for the 34 at 17-26 the parallel bar for the 35 at the 16-17 identifies the 36.

The parallel bar for the 36 to the 28-25 identifies the 37 on the parallel for the 20 to the 16-17; 30-32-36-34-37-33-30 defines the base perimeter of the fourth pylon; note that 30-39 is the axis of oblique symmetry, being 33-38 = 38-32 and 37-39 = 39-36; the solution process can use this way.

Extending the 16-20 and 17-39, you get the center 40 of the shooting range, out of bounds; to know the length 16-40, divide 16-17 into equal parts and, for example, let 16-41 be equal to a sixteenth of the 16-17.

The parallel for 41 to 17-40 identifies 42 by similarity; the length required in projection is 16 times 41-42 and in reality, 16 times 41-45, or 0-46; Graphically resolved, the aforementioned measurement is about 100 m.

The scholar is invited to reflect on the consequences of a different inclination of x' with respect to z, of a ratio $u_x' = u{:}n$ with n other than 5, of an initial straight line not parallel to x'.

It should be noted that on each new side of the regular axis polygonal of the viaduct, the deviation shows some significant variations of the new pylon; in a straight line, the base is rectangular with a misleading image, being, as already mentioned, 1-2 = 5u and 2-3 = 10u; In the deviation, the base of the pylon has six sides.

The opposing faces 5u wide are not parallel, the other opposing faces are parallel but not the same; the internal one is smaller than the external one; Having taken into account these properties has greatly facilitated the construction process of both the pylon and the roadway of constant width.

The separate figure was used to determine more quickly the carry-over of each successive angle, first in plane projection and then in oblique axonometry on xy; The reading is entrusted to the careful connection of the text with the drawing according to the progressive numbering.

The hidden parts are not suppressed but presented in a dotted line; the basic sections highlight the exposed parts in full line; the full-line head sections highlight the transformation of the head section and the relationship with the diagonal sections, also shown in full line, for transparency.

The viaduct is not sloping; reflect on the graphic consequences of the aforementioned diagonal sections in the presence of graphically appreciable slopes of constant value; Imagine this section guaranteeing a flat surface of the roadway.

The prismatic cavity of the caisson is marked only in the head section for the depth that can be perceived directly, indicated by the dash of the edge on the left at the bottom; The vertical and visible folds are part of the diagonal sections.

Reflect on the choice of the coordinated plan; With the same scale and available field (the sheet), it is not possible to represent the same viaduct length for an axonometric reference x, y, z'.

Every choice must be weighed; only in this way can an extended, communicative, rich image be offered, responding to the qualities of reliability and completeness of the idea.

The use of oblique axonometry, as well as any other projective pathway, allows us to illustrate a generic spatial operation; For example, you want to explain a way of creating an environment in a solid perspective.

If a consequent perspective of a given volume, seen from the point chosen as the center of projection, is perceived, without prejudice to the appearance of the volume, it can be conformed differently.

In the absence of significant clues, the observer will not be able to notice the mystification; sometimes the intervention is known and accepted for the invariance of the results; This is the case of bas-reliefs.

The scheme (Fig. 4.6) is based on a reference x, y′, z, C, with $u_y' = u$; the numbering follows step by step the development of the procedure, which, as it is repeated, must be suitable for making the transformation carried out imperceptible to the observer $V(V', V_{xy}')$.

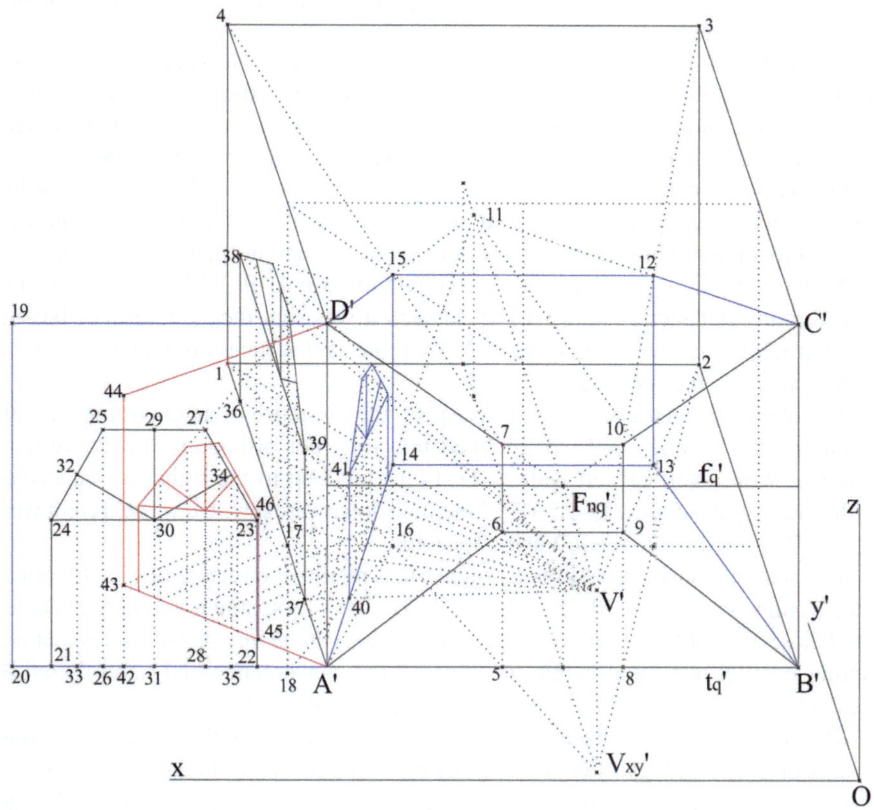

Fig. 4.6 Perspective illusion and space: interior environment with a trapezoidal floor A′-14-13-B′ inclined upwards, not orthogonal to the vertical painting, with a vertical left wall A′-D′-14-15 inclined to the right and trapezoid, with a trapezoidal ceiling D′-15-12-C′ inclined downwards, with a vertical right wall B′-13-12-C′ inclined to the left and trapezoid

The perspective frame A′-B′-C′-D′, proscenium with respect to V of the parallelepiped environment having the base A′-1-2-B′ belonging to xy′, and the bottom 1-2-3-4 parallel and equal to the proscenium, is parallel to xz.

The perspective from V is defined by F_{nq}', the vanishing point of all the lines orthogonal to the frame, and by the distance V′-F_{nq}'; the plane A′-1-2-B′ is defined by the straight line t_q' and the escape f_q' for F_{nq}'; the axonometric projection of the orthogonal projection of V on xy is V_{xy}'.

The perspective of all the edges of the parallelepiped is on the picture A′-B′-C′-D′; that of A′-1 is A′-F_{nq}'; once V_{xy}'-1 has been drawn, the 5 is obtained; that of the vertical plane for F_{nq}'e for 1-4 is the vertical for 5; on it the V′-1 and V′-4 identify the 6, perspective of 1 and the 10 perspectives of the 4; similarly, 8, 9 and 10 are obtained.

The perspective of the backdrop is the rectangle 6-7-10-9-6; these vertices are connected with those of the proscenium belonging to the painting; V "sees", if he stands still, every point of space in the corresponding perspective point on the picture that covers the real one.

If you draw it on the painting, V perceives it and places it where he has "seen" it; A′-6 is A′-1 and from A′ to 6 he has the sensation of proceeding away from A′ to 1; the trapezoid A′-6-9-B′ does not understand it as such but as a basic rectangle A′-1-2-B′ and understands it, by habitual knowledge, as a horizontal plane.

For the same reason that he cannot imagine a non-horizontal floor, V, from his consolidated experience, believes that he is observing a parallelepiped environment with orthogonal or parallel faces, naturally, regularly, and normally rectangular.

Now on the V′-F_{nq}' the vertex of a pyramid of base A′-B′-C′-D′ is placed in 11; consider any flat section 12-13-14-15 in such a way that the said vertices belong respectively to the rays V′-3, V′-2, V′-1, V′-4; the said vertices are seen by V on the picture again in 10, 9, 6, 7, as 3, 2, 1, 4.

The observer cannot perceive, nor suspect, the existence of an environment with a trapezoidal floor A′-14-13-B′ inclined upwards, not orthogonal to the vertical painting, with a vertical left wall A′-D′-14-15 inclined to the right and trapezoid, with a trapezoidal ceiling D′-15-12-C′ inclined downwards, with a vertical right wall B′-13-12-C′ inclined to the left and trapezoid.

The section considered forms a volume that retains the rectangular and vertical background plane 12-13-14-15, linked to A′-B′-C′-D′ by spatial homothety of center 11; The section of the pyramid, by generic choice, can be on opposite sides that are neither parallel nor equal, while maintaining the convergence in 11 and therefore also the direction of maximum slope with respect to the frame.

The section plane in this case has a trace on the picture with the geometric role of axis of the homology of center 11; The new volume is not thought of, because it is unusual, unimaginable, atypical in normal experiences, perfectly replaced by the thinkable one, which is much deeper.

To a careful eye, the artifice would be trivially unmasked by a small movement of V; the useful effect of the problem lies in the fact that, if one does not have the depth A′-1 to place everything that accompanies, animates and enriches the environment of that depth, the mystification of the lesser depth and conformation makes it effectively possible from that and only position of V.

The depth A′-17 in the y′ direction, used for the section plane, is obtained, considering 14-16 and 16-17; A significant example is appropriate.

On the left wall, draw a door, giving its scheme on the wall overturned on the plane of the painting; Being parallel to XZ, this scheme is shown in true form and size at the U scale.

The D′-4 is reversed into D′-19; the rectangle D′-19-20-A′ is derived; A′-36 and 36-38 are equal to A′ 21 and 36-38; for all the other vertices, proceed in the same way, obtaining the oblique axonometric scheme on the wall A′-1-4-D′.

Note that V′-1 intercepts 14, corresponding to 1, obtaining in 6 the perspective of both; the 37-V′ on the A′-14, corresponding to A′-1, intercepts the 40, corresponding to the 37; the relative and only perspective is omitted.

The 37-39 is projected into 40-41; proceed in the same way for all vertices and arrive at the drawing of the door on the wall A′-D′-15-14; the perspective, from V on the painting, of both schemes is unique; the anomaly introduced is not perceived by V.

The true form of the second scheme is obtained by reversing the painting, by rotation around A′-D′, the A′-D′-15-14-A′; 17-A′ is reported to 17-18; 16-18 is the true length of A′-16 and is brought back to A′-42.

The 16-14 is back in 42-43, and A′-43 is the overturning of A′-14; the 14-15 is brought back to 43-44, and the 44 is joined with D′; the side wall A′-D′-15-14 in true shape and size is the A′-43-44-D′.

D′-15 is D′-44, while 15-12 is in true magnitude, and 12-C′ is equal to D′-44; the floor A′-14-13-B′ can also be constructed, since 14-13 and A′-B′ are in true size, and A′-14, equal to B′-13, is equal to A′-43.

The true shape and size of the right wall C′-B′-13-12-C′ is equal to the left wall, i.e., it is equal to symmetrical to D′-A′-43-44, since the vertical plane V_{xy}'-V′-F_{nq}'-11 is of symmetry for the parallelepiped volume from the proscenium to the bottom.

In the absence of such symmetry, for the trapezoidal wall C′-B′-13-12-C′, we must proceed, obtaining it as D′-A′-36-38-D′, we arrived at D′-A′-43-44-D′.

For the transfer of the vertices of the door, such as 40, belonging to A′-14, the parallel to 14-43 is conducted for 40, identifying A′-43 as 45; 40-41 is back in 45-46; all vertices can be obtained similarly; that is, the oblique affine homology can be used, having the axis in A′-D′ and a pair of corresponding points 14 and 43.

The scholar is left with the burden of thinking and designing a coffee table with a rectangular and horizontal top and with all four legs, as is normally the case, of equal height, resting it on the horizontal floor x-y′, with one side parallel to x.

The scholar draws on A′-6-9-B′-A′ the perspective projection from V′, after reversing the rectangular floor having one side A′-B′ and the other A′-20; it must proceed as for the door on the left wall.

Then, after reversal defined by the trapezoid having the bases parallel to A′-B′ and 14-13, and the oblique sides equal to A′-43, the projection of the base is obtained on A′-14-13-B′, which is no longer rectangular and, however, prospectively coinciding with the one already drawn.

Bring the shelf to height; it is verified that the heights of the four legs are two by two, equal in the x direction and different in the other direction; returning the projection on the walls, we find that the table is lopsided and the shelf is a trapezoid, with the major and minor bases parallel to x; moreover, the distance between the said bases is less than the corresponding one in the unaltered table.

In this way, we have the opportunity, having all the dimensions of the illusory environment, to build it and experience that visual perception is misled; the shortened environment is useful, for example, in bas-relief, having to necessarily pretend to have a greater depth of field.

It is reiterated that the immobile position of V must correspond to that used in the drawing, V'-F_{nq}', at scale u, and that, to avoid external elements revealing the contraption, it is advisable to enclose the room in a box and let the view be spied through an eyepiece, unaware of the conditioning.

In the case of bas-relief, it must be said that in practice, the attention is not so much pushed on the thread of verifying the correct geometry, which, on the contrary, is the basis of the geometric experience illustrated.

And again, in the bas-relief on a column with a quadric surface, the deformity is not highlighted, or is neglected, involved and satisfied by prevailing cultural, historical, artistic, and material factors, in a more humanistic than scientific reading.

Returning to the example, it remains to point out the possibility that the illusion is applied not to the trivial parallelepiped volume and to the pyramidal volume, which is equally trivial; It can be any environment that you want to make look like a completely different environment.

Evidently, the difficulties of the consequent translation are much more demanding, at great risk, requiring a much wider graphic field and a graphic scale closer to the truth.

Note the different conception between this exercise and the previous one, that of the viaduct; In the exercise presented here, the field of action is, in reality, in the solid model and not in the graphic representation of the model.

Therefore, the interest in determining the deformed model so that it seems to be the one that can be thought of and perceptually correctly responds to expectations prevails.

Regular Polyhedra

The five regular polyhedra have already been discussed in other chapters, reflecting and clarifying their geometric properties and activating the inspiration of possible applications, both of whole volumes and of their sections, intersections, combinations, etc.

Here we want to draw (Fig. 4.7) a semi-dodecahedron in oblique axonometry referring to the triplet x, y, z′, O, with $u_z' = u$ and seen from below.

In particular, the pentagonal face of the roof belonging to xy is associated with the development on xy of the remaining faces; the projections are aggregated below

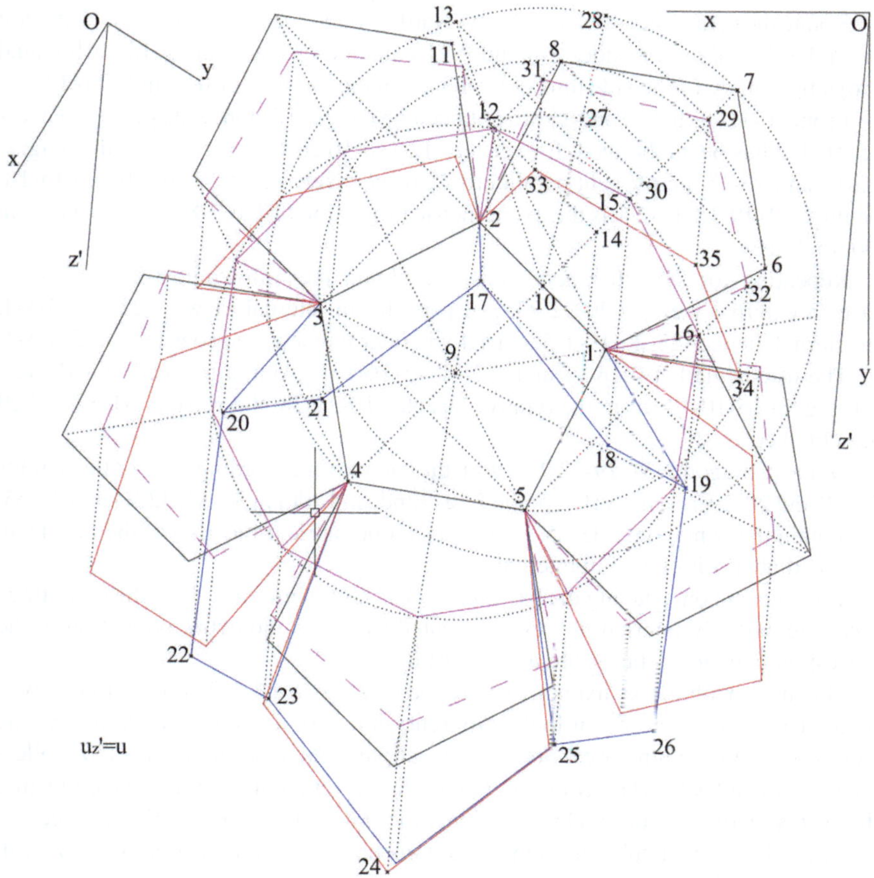

Fig. 4.7 Representation of a semi-dodecahedron in oblique axonometry referring to the triplet x, y, z′, O, with $u_z{}' = u$ and seen from below

the aforementioned face, concluding the volume; This is followed by the drawing of each of the faces in the middle position of the rotation that proceeds from the flat development to the volume.

Geometry is discussed net of the designation of points in space by means of coordinates; Therefore, it does not seem a provocation if different triplets can be given to the reference, without prejudice to the result of the search, always assigning to z′ the same direction z′ with $u_z{}' = u$.

Given on xy the pentagon of base 1-2-3-4-5 in true size, by symmetry with respect to the side 1-2 the pentagon 1-6-7-8-2-1 is obtained; this is true for the other pentagons; there is the development on xy of the surface of the semi-dodecahedron in six pentagons; the center is 9 and the average of 1-2 is 10.

Rotate the pentagon 1-6-7-8-2-1 downwards around 1-2 and the adjacent pentagon around 2-3, making the sides 2-8 and 2-11 coincide in space; the 2-12, orthogonal projection on xy of the side thus obtained, is on the bisector of the angle 11-2-8.

From 10-12-13 the 14 and 15 are obtained on the 9-10-7; Once the pentagon 2-12-15-16-1-2 has been obtained, for vertices 12 and 16 and for vertices 15 the heights 12-14 and 13-15 are brought parallel to Z′, respectively in 12-17, 16-19 and 15-18; pentagon 1-19-18-17-2-1 is the axonometric projection of 1-6-7-8-2-1; it is a face in external view.

Repeating for the other pentagons, we obtain the following: 3-20-21-17-2-3, face in external view; 3-20-22-23-4-3, partially in internal view; 4-23-24-25-5-4, in internal view; 5-25-26-19-1-5, partially in internal view, and so is the 1-2-3-4-5-1.

The true angle between xy and 1-2-12-15-16-1 is given, after section and overturning, in 15-10-13; from the relative bisector 10-27-28, the vertices 31-32 and 29 are obtained.

Of the pentagon 2-31-29-32-1-2, with the carry-over according to z′, of the heights 27-30 and 28-29, respectively, for the 31st, 32nd, and 29th in 31-33, 32-34 and 29-35, we obtain the pentagon 2-33-35-34-1-2, an axonometric projection of the face in an intermediate position and in internal view.

The adjacent one, having the base side 2-3, is also in internal view; the other three faces are partially internal in view, prevented by the axonometric projection of the corresponding closed faces, already determined.

The angle with the xy axonometric frame of 2-33-35-34-1-2 appears to be given by 35-10-29; for each face it looks different; in reality it is equal to 29-10-28; one can make use of symmetries, homologies, alignments, etc.; the circles that collect the vertices and parallel to xy in axonometry move, remaining parallel and collecting the corresponding vertices; On reflection, other interesting singularities emerge.

Draw (Fig. 4.8) a cube in oblique axonometry, in the reference x′, y, z, O, with $u_x' = u$; it would seem a pretentious, banal, too easy request; It's not like that!

It depends on the inclination of the x-axis′, 45° projection of the x-axis; it also depends on the position of one of the faces and the direction of one of the sides.

It is understood that the choice has the purpose of a challenge; knowing that we never know enough, we must reject the clichés of the classic example of the cube that appears as a hexagon.

A teacher, "not by chance," must train on the minefields of demanding problems and know how to solve them with reflective balance at a high level of wisdom.

The support surface is xy; it overturns on yz by rotating on y; the basic square is drawn, and the corresponding projection of each vertex is obtained; a rectangle appears; reporting the height equal to the base side for each vertex results in the visual perception of a parallelogram.

By suitably joining the averages of some sides, the hexagonal plane section A′-B′-C′-D′-E′-F′-A′ is obtained, which is regular in space; its center is the center of cube G.

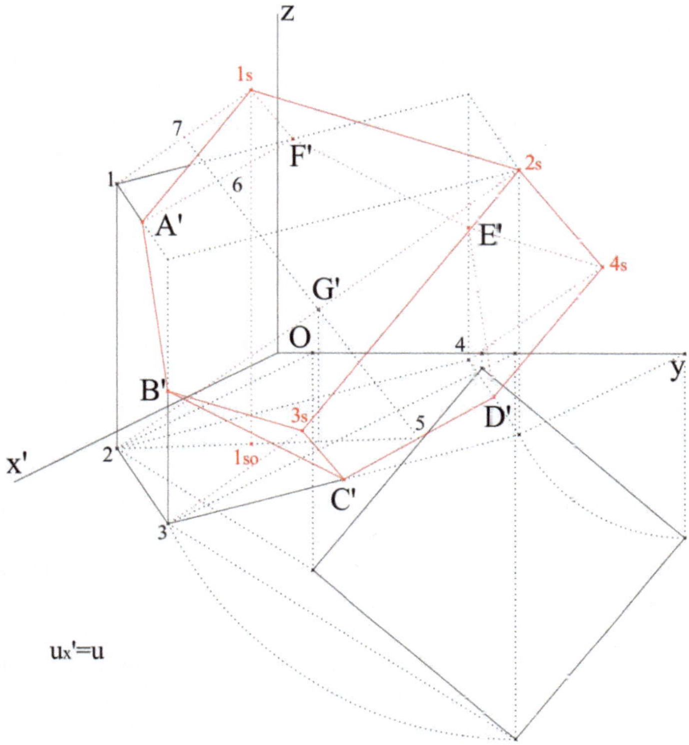

Fig. 4.8 Oblique axonometry of a cube

We want to consider only the underlying part and aggregate them, in orthogonal symmetry with respect to the plane of the hexagon, the sectioned volume; the orthogonal direction is the diagonal of the cube 2-G′-2_s; the 2_s is the symmetry of the vertex 2.

The vertices of the hexagon are joined to their respective symmetricals; the median of the 5-G′-6 hexagon is in the 1-2-5-D′ plane; the 7-1_s is symmetrical with the 1-7; the sides 1-A′, 1-F′, and 1-2, converging in 1, have as symmetrical, respectively, 1_s-A′, 1_s-F′, and 1_s-2_s.

A′-1_s-F′ is not in sight; 1_s-A′ limits the view of 1-F′; B′-3, parallel to 1-2 has for symmetric B′-3_s, parallel to 1_s-2_s, and 3-3_s is parallel to 2-G′-2_s; the B′-C′-3_s is in sight; at 2-4, parallel to 1-F′ is symmetrical 2_s-4_s, parallel to 1_s-F′.

In conclusion, in the symmetrical volume, the faces A′-1_s-F′, F′-1_s-2_s-4_s-E′, E′-4_s-D′ are not visible; instead, the D′-4_s-2_s-3_s-C′, symmetrical of the beveled base, appears in view, as well as the C′-3_s-B′ and the B′-3_s-2_s-1_s-A′; the latter appears immeasurable.

Observing carefully and with a lot of reflection, one perceives the original volume, the symmetrical volume, the symmetry correlations, the projective deformations, including the persistence or opposition of the exposure of the various faces.

The symmetrical body aggregated through the hexagon to the original one can, in turn, undergo a rotation that preserves the center G′ and brings B′ to A′, C′ to B′, etc.; between the two bodies in contact, there is no longer a link of symmetry.

In representing this new hypothesis, it is understood that everything changes, positions, directions, shape, size, solution process, exposures of faces, etc.; the odds 1_{so}-1_s change position and value, and so does 2_s-D′, etc.; a challenge that is left to the scholar.

The examples proposed must not constitute occasional applications, but aimed at stimulating investigations, reflections, curiosities, doubts, clarifications, conclusions, new hypotheses of in-depth analysis, to guarantee the principle, repeatedly maintained, of committing oneself to new conquests of knowledge, without end.

All this, open to all, can have a good outcome for those who have matured step by step, both for the expanding intuition and for the natural selection of the most gifted, the awareness of the relationship between the cost of serious study and the profit of valid knowledge.

The new trial shifts the attention to the representation of an icosahedron in oblique axonometry entrusted to x′, y, z, O, with $u_x' = u$; One diagonal is vertical, and the XY reference plane contains the pentagon of the vertices of the five first-level triangles competing at the lowest extreme vertex of the vertical diagonal.

The projection of the five equilateral triangles of side 1-2, inscribed in the circle 7 of center 6, is drawn on (x, y), which is also the orthogonal projection on xy of the third vertex common to the aforementioned five triangles (Fig. 4.9).

The isosceles triangle 2-3-6 represents the equilateral 2-3-6* inclined on (xy) so that on the orthogonal for the 6 to the 8-6 the rotation of the 8-6* intercepts the 6-9*; The 6-9* = 6′-9′ is the difference in altitude between the pentagon and the lowest vertex of the icosahedron.

Combined 9′ with 1′, 2′, 3′, 4′, 5′, we obtain the axonometric projection of the five triangles, constituting the part of the volume at negative altitude; from the projection on (xy) of the five vertices 10, 11, 12, 13, 14, of the upper pentagon of the volume, we obtain on xy at altitude 0 the points $10_o'$, $11_o'$, $12_o'$, $13_o'$, $14_o'$.

In orthogonal projection on (xy), the triangle 2-3-6 is the adjacent and highest triangle 2-3-10 and the odds of 10 are obtained in 10-15*, being 8-15* = 8-6*; 10-15* is brought back to 6′-15′; Lifting the vertices 10′, 11′, 12′, 13′, 14′ are obtained.

The middle of 6′-15′ is the 16′, projection of the center of the icosahedron; bring back the 6-9* in 15′-17′; the 17′ is the axonometric projection of the highest vertex and must be joined with the vertices 10′, 11′, 12′, 13′, 14′, constituting the highest part of the volume, as opposed to the previous lower one, with respect to which it appears rotated.

By joining the vertices of the first horizontal pentagon with the one now obtained, 10 triangles are obtained; this completes the projection of the icosahedral volume; with attention and rational perception, and not just visual, it is possible to highlight the triangular faces in view.

Taking into account that the sides of the upper pentagon are visible 14′-10′, 10′-11′, 11′-12′, and of the lower pentagon the sides 2′-3′, 3′-4′, the triangles 17′-14′-10′, 17′-10′-11′, 17′-11′-12′ are visible, then the triangles 9′-2′-3′ and 9′-3′-4′ and, of the

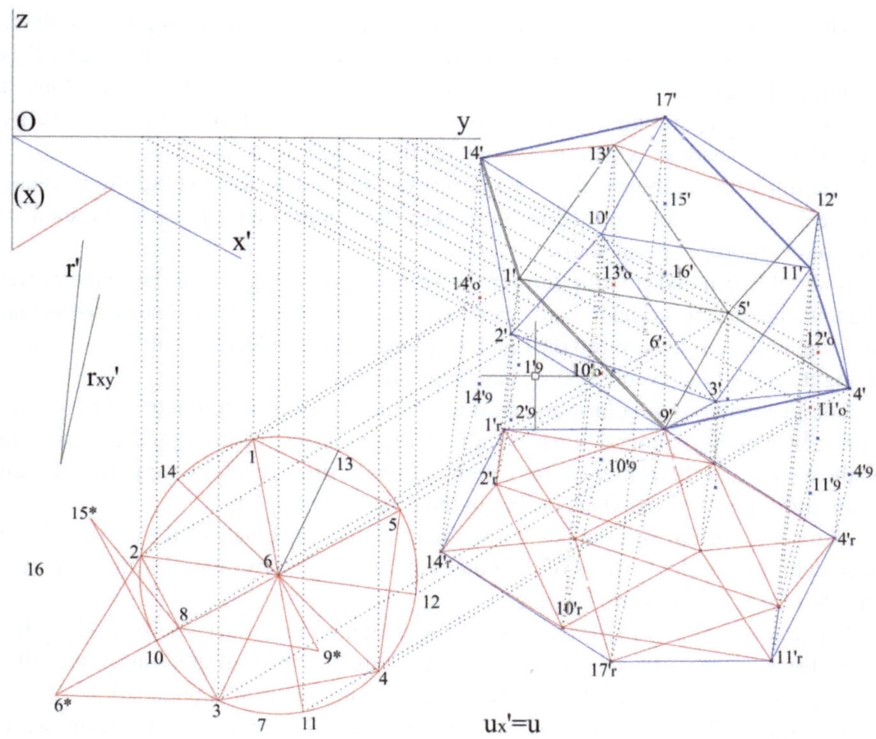

Fig. 4.9 Representation of a hicosahedric volume

central band, the triangles 14′-10′-2′, 2′-3′-10′, 10′-11′-3′, 3′-4′-11′, 11′-12′-4′, for a total of ten faces; the remaining ten are hidden.

Therefore, looking in the direction of the oblique projection, the apparent outline of the icosahedron is given by the spatial polygonal 9′-2′-14′-17′-12′-4′-9′; The apparent outline of the same volume seen according to Z, i.e., according to 17′-9′, is given, on the other hand, by 14′-2′-10′-3′-11′-4′-12′-5′-13′-1′-14′.

It seems useful to propose to read the apparent contours according to a given direction; For this purpose, given the direction r(r′, r_{xy}'), we consider a plane that does not project r, for example, the horizontal plane that contains the 9′.

On this, we obtain the vertical projections $1'_9$, etc.; for $1'_9$ we lead the parallel to the r_{xy}' until we intersect the parallel to the r′ conducted for 1′ in $1'_r$; repeating, we join the ones that are the outermost, constituting the polygonal $1'_r$-$14'_r$-$17'_r$-$11'_r$-$4'_r$-9′-$1'_r$, with respect to which no top projection is external to it.

From this, we go back to the corresponding sides, identifying the required apparent outline on the volume in 1′-14′-17′-11′-4′-9′-1′; projections of vertices clearly extraneous to the outline, such as the 2′, can be avoided; Verification is carried out for the tenth minute.

Therefore, the aforementioned line separates the faces seen looking from infinity in the direction r from the remaining hidden; If the improper point indicated by R is a source of energy, obtaining the aforementioned polygonal of apparent contour, it can be called a separating line of shadow, since it divides the faces exposed to energy from those in shadow, which do not receive the energy used.

Dealing with the theory of incident energy in the future, the contour on the plane for the $9'$ will be called the shadow brought by the volume on the plane of the shadow due to the improper point source.

It should be noted that the farther the vertex moves away from the plane of the shadow, the farther its shadow moves away from the volume; so, it also happens that the smaller the angle between the r and the plane of the shadow, the more the shadow brought on the plane is elongated, out of the sheet.

The problems on regular polyhedra are endless; the volume under investigation can be broken up, exploded into pyramids all the same, having the equilateral triangle as their base and the center of the volume as their vertex; each of the pyramids can be thought of as translated according to the direction of height.

The center of the volume is symmetrical, so that, once half of the vertices have been determined, the vertices of the other half are located by central symmetry; it is easy to obtain a concentric volume.

It is easy to obtain the projection of the center of each face by identifying the projections of the vertices of the dodecahedron, dual of the icosahedron; conversely, once the dodecahedron has been drawn, the dual icosahedron is obtained.

The image of the projection of the icosahedron, carefully observed in its network of edges for transparency, reveals a series of specular symmetries, useful not only for simplifying its determination but also for possible and valuable checks of the results.

The copiousness of reasoned applications creates in the scholar the shrewdness, the sense of spontaneous, present, growing, reassuring verification, freeing the little exercised from the considerable perplexities and entanglements of which, inevitably, the little exercised is a victim.

Re-examining from the starting point, having placed the volume in non-independent equilibrium, resting on a vertex, has led to a process, hopefully, now digested; No one forbids considering the same volume resting on a face or on an edge.

In such cases, the hypothesis can be accepted as an unavoidable suggestion to find oneself a winner on the consequent different, demanding, formative resolution procedures.

It should also be noted that in the process illustrated, the use of exclusive and repeatedly applied routes was insisted upon; To achieve gradual progress, one can follow, in truth, all the available paths.

Randomness, opportunity, reasoning, simplification, the deliberately different approach, must guide the scholar in the free research activity to achieve the result; There is no single solution, there is no mandatory path, no method to memorize.

Here, one of the purposes of this singular treatment of learning is to always and only recommend reasoning, a guarantee of reliable acquisition of answers to the "whys", to aspire to a laboriously but correctly progressive personal baggage, based on the geometry of the drawing.

Semi-regular Polyhedra

At the vertex of one of the five regular polyhedra converge all equal edges of faces all equal, regular polygons of a single type; At the vertex, the sum of the angular values of these faces must be less than 360° and the faces converging to the common vertex cannot be less than three, otherwise the solid angle of the volume is not determined.

This constraint limits the number of regular polyhedra; with squares, there is only one polyhedron, the hexahedron or cube; with the equilateral triangle, three can be formed, the tetrahedron, the octahedron, the icosahedron, with the regular pentagon, there is only the dodecahedron.

Regular polyhedra have a center; the distance from the vertices has only one value, from the averages of the edges a unique and smaller value, and from the centers of the faces a value less than the above.

It follows that regular polyhedra admit a circumscribed sphere containing all the vertices, an intermediate sphere touching the edges at the midpoint, and an inscribed sphere touching all the faces in their center.

Such volumes can have many properties, such as convexity, polarity, assembly capacity, parallelisms or orthogonality of faces or edges, faces in translated opposition or in central symmetry, axes or planes of orthogonal or oblique symmetry.

Euler's formula "number of faces + number of vertices = edges + 2" binds the three elements and is valid for any polyhedron, regular, irregular, convex, or concave.

Smooth the three squares competing to a vertex of a cube with a plane that halves the edges; each of the eight vertices removed is replaced by an equilateral triangle.

A cuboctahedron is generated, with six squares and eight triangles as faces; Faces are regular polygons with a single edge value but no longer of a single type, but of two types, and is called a semi-regular polyhedron.

The group of polygons converging at the vertex of the volume thus modified is composed of two squares and two triangles; this genesis can be thought of not by chamfering, but by directly associating solid angles consisting of two squares and two equilateral triangles; The angular sum is 300° < 360°.

A flat chamfer with removal of the vertex can produce the doubling of the sides of the original faces, and the polyhedron retains the name followed by the word "trunk".

If the six square faces of the cube are transformed into six regular octagons, eight equilateral triangles must be replaced by the eight vertices removed; at the top, there are two octagons and a triangle, for a total of 330°; the volume generated is the truncated cube.

There are 13 possible group couplings forming a trihedron angle; from the cube or octahedron, its dual, the cuboctahedron is generated; from the dodecahedron and the icosahedron, its dual, the icosidodecahedron is obtained.

By truncating regular polyhedra, the truncated tetrahedron, the truncated cube, the truncated octahedron, the truncated dodecahedron, and the truncated icosahedron are obtained; By truncating the semi-regular polyhedra, already presented, the truncated cuboctahedron and the truncated icosidodecahedron are obtained.

Associating three squares and an equilateral triangle at the vertex, we obtain the rhombicuboctahedron; By associating three types of regular polygons, i.e., a regular pentagon, two squares, and an equilateral triangle, the rhombicosidodecahedron is obtained.

Associating three equilateral triangles and a square to the vertex gives the cubic snub; Associating four equilateral triangles and a regular pentagon gives the snub dodecahedron.

In semi-regular polyhedra, we can consider both the circumscribed sphere, which contains all the vertices, and the one that contains the midpoints of the edges, being equal chords; the faces are not all the same and their centers are not all equidistant from the center of the semi-regular polyhedron; the inscribed sphere cannot be traced.

The five regular polyhedra are called Platonic, while the 13 semi-regular polyhedra are called Archimedeans; It is possible to assemble some regular polyhedra with different semiregulars and even semiregulars.

Examples of assemblages between two different semi-regular polyhedra: tetrahedron and truncated tetrahedron; tetrahedron and octahedron; cube and octahedron truncated, octahedron and cuboctahedron; octahedron and truncated cube.

Examples of assemblages between three different semi-regular polyhedra: cube, tetrahedron, and rhombicuboctahedron; cube, cuboctahedron, and rhombicuboctahedron; cube, truncated octahedron, and truncated cuboctahedron; truncated tetrahedron, cuboctahedron, and truncated octahedron; Truncated tetrahedron, truncated cube, and truncated cuboctahedron.

From the above, it is clear that for convinced and convincing graphic applications, considerable experience on these polyhedra is needed; their representation must also be in-depth for any possible significant fractionation.

Want the oblique axonometry of a truncated cuboctahedron (Fig. 4.10), referring to x, y, z′, O, with $u_z' = u$, according to the genesis that derives it from a cuboctahedron, obtained from a cube on the given side and freely resting on xy.

Consider the six square faces of the cube, the six dual squares, having as vertices the midpoints of the sides of the cube; these resulting squares enclose eight equilateral triangles; By eliminating the pyramids at the vertices, the cuboctahedron is obtained, consisting of 14 faces, 12 vertices, and 24 edges.

Let each square be transformed into a regular octagon; it follows that each triangle becomes a regular hexagon; 12 squares are created between octagons and hexagons; the truncated cuboctahedron is obtained, having 26 faces, 48 vertices, and 72 edges; 13 faces are visible; the other 13 not shown are not marked.

Drawing is not enough; you have to observe the results; for example, the squares of the truncated cuboctahedron oppose the edges of the generating cube; obtained the vertices of the half, and all the other vertices are obtained by symmetry with respect to C′.

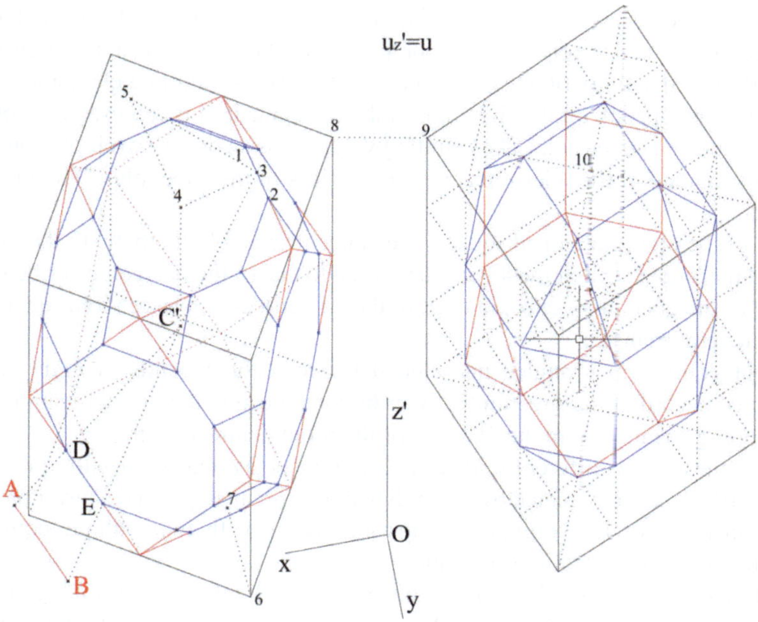

Fig. 4.10 Oblique axonometry of a truncated cuboctahedron (on the left side) and of a rhombicuboctahedron (on the right side)

For a given length A-B of the edge of the truncated cuboctahedron, a side D-E belonging to xy is projected from C′, and the position of the side A-B is obtained; by homothety of center C′, the correspondents of all sides are drawn, and the desired truncated cuboctahedron is obtained.

Note that the 1-3-2 side, its 4-3 axis and C′-4 are in true magnitude; the 5-3, obtained by placing the 4-C′ in 4-5, orthogonal to the 4-3, is the radius of the intermediate sphere, the true dimension of C′-3; similarly, the true dimension of C′-2, the radius of the sphere circumscribed by the polyhedron, is obtained from C′-4-2.

Countless dimensional and relative position problems can easily be solved; for example, the distance 6-7 can be obtained by proportionality with the diagonal of the original cube; 6-7 is also orthogonal to the position of the hexagon.

Similarly, it is possible to obtain the distance between a square and the edge of the original cube opposed to it; The projection on XY of the vertical square is considered.

It is interesting to imagine prismatic holes with a square or hexagonal, or octagonal direction and their intersections within the volume; generic and particular section planes can be made, parallel to a face, to an edge, at a given distance from them or from the center of the volume.

One can think of a fragmentation and imaginative combinations of parts, with direct interventions on the complete image; the possession of space is mature when the inspiration is satisfied by decisive and promotional responses.

The more extensive the usability is, the further away we are from the habit of reduced, trivial, and poorly formative preparations; it is a serious choice, appropriate to the dignified activity of a teacher, and which suits his unfailing charisma.

In the same panel, on the right, a rhombicuboctahedron is represented, obtained from a cube on a given side; The resulting volume has a convex envelope, consisting of eight equilateral triangles replacing the vertices of the cube, plus 18 intercalated squares.

Six of the aforementioned squares are concentric and homothetic to the faces of the original cube; each of the 48 edges is equal to that of the octagon inscribed in the square of the generating cube; at each of the 24 vertices converge three squares and an equilateral triangle.

Eight equilateral triangles intercluded between three bands of eight squares combined according to three mutually orthogonal positions, after unification of the overlapping squares, form the 26 faces of the rhombicuboctahedron.

The genesis, read differently, identifies a regular octagonal prism with square faces and a vertical axis, completed by two polyhedral caps, an upper and a lower pavilion, each consisting of four oblique squares alternating with four oblique equilateral triangles and a horizontal closing square.

Rational perceptual observation helps to highlight the possible generative processes, as well as the significant decompositions and singular and freely invented compositions, an indication of developed and attentive personal creativity.

Visual analysis confirms the 13 faces in view, suggests parallelisms and orthogonality between the faces and the edges, central symmetry of faces, vertices, and edges, axes and planes of symmetry.

Reflect on the impossibility of central symmetry planes containing an edge; the choice of appropriate data ensures very simplified views, or with carefully accessible polygonal interweaving structures; it is useful to obtain the radius of the circumscribed sphere containing the vertices and of the one containing the midpoints of all the edges.

A spatial reading notation: with respect to the reference marked, each of the two generating cubes has a coplanar base on xy, and the alignment of two edges with the frontal segment 8-9 has been highlighted for both; the two bases, however, are rotated, and the 4-10 segment of the ends of the axles is not frontal.

One of the very challenging semi-regular polyhedra is the dodecahedron snub, consisting of 12 regular pentagons and 80 equilateral triangles; at each of the 20 vertices of the original dodecahedron four triangles converge; there is no contact between the pentagons and they are moved away, as if by explosion from the center with slight rotation; position and central symmetry are preserved; the vertices are 60 and the edges 150.

Of the plane development on xy, consider the regular pentagon of center 0 and vertices 1-2-3-4-5, the equilateral triangles 1-2-6, 1-6-7, 1-5-8, and 1-8-9, with 0-1 axis of symmetry.

Rotate the coplanar triangles 1-2-6-7-1 around 1-2 and 1-5-8-9-1 around 1-5 temporarily and rigidly, until the 9 joins with the 7 in 10; 6-7 translates to 10-11, 8-9 to 12-10; 1-2-11.

The triangles 1-2-6 and 1-6-7, integral and coplanar, rotate around 1-2; the 6-7, parallel to 1-2, rises above xy, describing a straight circular cylindrical surface of axis 1-2; the 6 rotates with radius 11-6 around 10, average of 1-2, describes a vertical circumference in space, of track 10-6; projection 6 moves towards 10, on 10-6.

Similarly, the triangles 5-1-8 and 8-1-9, integral and coplanar, rotate around 5-1; the 8-9, parallel to 5-1, rises to xy, describing a straight circular cylindrical surface of axis 5-1; the 8 rotates with radius 11-8 = 10-6 around 11, average of 5-1, describes a vertical circumference in space, of track 11-8; projection 8 moves to 11 on 8-11.

The sides 6-7 and 8-9 move and join in 12 on 0-1; they are united in 12, with their triangles in the conditions of coplanarity; The projection on XY of the triangles 1-2-6 and 5-1-8 is given by the triangles 1-2-13 and 5-1-14.

For the fold around 1-13 and 1-14, the 12 runs along 1-12, towards 1; the five vertices, such as 13 and 14 are associated with the five vertices, such as 15, a position reached by 12; in projection, a regular decagon is obtained.

The basic pentagon appears surrounded by the projection of ten triangles; vertices such as 13 have the quota 13-16, being 10-16 = 10-6; vertices such as 15 have odds of 15-17 being 1-17 = 1-6.

This completes the first significant part of the dodecahedral snub, having the polyhedral shape of a bowl, or, if you like, an upside-down pavilion.

Now, to the development of the first initial module, consisting of a regular pentagon and four equilateral triangles converging at vertex 1, the pentagon 6-7-18-19-20 and the triangles 6-20-21 and 6-21-22 are added.

In the new module, the four triangles converging at vertex 6 are considered; of 6-7-1, we have the projection 13-15-1; of 6-1-2, we have the 13-2-1 projection; of 6-21-22, we have the projection 13-23-2; it remains to identify the corresponding 20th.

Obtain in 24 the projection of the center of the triangle 1-2-13 and in 25 its position on the 10-16 of the vertical plane for 4-10-13 overturned; the orthogonal for 25 at 10-16 intersects in 26 the orthogonal in 0 at 4-10.

The 26-16 = 26-4 is equal to the radius of the sphere circumscribed by the polyhedron; the orthogonal projection on xy of the center of this sphere is 0 and its dimension on xy is equal to 0-26; It is possible to construct the projection of the upper pentagon and the triangles attached to its vertices, operating by central symmetry.

In projection, the sides correspond to the vertices of the pentagon, and the relative design rotates 36°; For example, vertex 2 of 2-3 moves to 28, 27 to 29, 3 to 30.

Of the pentagon 7-18-19-20-6, the 6-7 is positioned in orthogonal projection on xy in 13-15; for symmetry, the 18 moves along the 1-15-12, trace of the vertical plane, which contains the center; Flip this plane over on XY

The center C of the circumscribed sphere is set to 31; the circumference of center 31 and radius equal to 26-16, distance between the center and the spatial vertex corresponding to 18, intersects the circumference of center 17 and radius equal to 6-7 in 32, which acts as an inverted 18.

The orthogonal for 32 to 1-15 obtains 33, the position of the projection of the spatial 18; 33-32 is the odds of 18 with respect to xy; consider for the 33rd the parallel to 15-13, corresponding to the 18-20; the 34-35-19, for proportionality, is carried over to 36-37-40; 40 corresponds to 19; You can mark the pentagon 13-15-33-40-6.

The projection of the triangle 6-20-21 is placed in 13-6-23.

Parabolic Point Quadrics

Let us represent in oblique axonometry a quadric with parabolic points, which by definition is that surface defined by the set of infinite rectilinear generators of a bundle, with an improper or proper center, each of which has in common a single point, proper and exclusive, of the infinite points of a directrix, which, if it is flat, is always and only a conic.

A tangent plane, such as a surface has in common with it always and only a generating line; the surface remains all on one side with respect to the tangent plane; if the center of the beam of the generators is improper, it is a cylinder.

If the aforementioned center is proper, it is the vertex of the quadric, with two pitches, and it is a cone; any flat section can be taken as a guideline; If the position of the datum leader is orthogonal to the axis, the cylinder or cone is right.

Consider a cylinder (Fig. 4.11); the reference is x, y, z′, O, with $u_z' = u$; the cylinder is straight, the trace on the generic plane $\alpha(t_{\alpha xy}, t_{\alpha xz}', t_{\alpha yz}')$ is the ellipse is of axes a, b.

To represent on $\pi = xy$ the projection σ_α' of the basic ellipse σ_α belonging to the plane α you have to flip α on π, rotating it around the relative trace on π, $T_{\alpha x}$-$T_{\alpha y}$.

Just rotate the true distance 1-$T_{\alpha z}$ around 1; this is obtained by plotting the O-1 orthogonal to the $T_{\alpha x}$-$T_{\alpha y}$ for O; the true distance 1-$T_{\alpha z}$ is obtained as a hypotenuse on the cathetes O-1 and O-($T_{\alpha z}$), being O-($T_{\alpha z}$) = O-$T_{\alpha z}'$; it is reset from 1 on O-1 to O-$T_{\alpha z}$*.

The $T_{\alpha x}$-$T_{\alpha y}$-$T_{\alpha z}$* triangle is the true shape and size of the triangle of α, of $T_{\alpha x}$-$T_{\alpha y}$-$T_{\alpha z}'$ projection; in it, the pair of axes of σ_α, the major a*, the minor b*,

Of the pair a*, b*, the projection a′, b′ is obtained, reporting by proportion on the corresponding sides the points of intersection of the pair with the sides of the overturned triangle, without prejudice to those with the $T_{\alpha x}$-$T_{\alpha y}$ side, track, and axis of rotation of the overturning.

From the conjugate diameters a′, b′, the axes are obtained and therefore the basic ellipse σ_α'; the randomness of the quasi-coincidence of the diameters conjugated to the axes is felt.

To carry-over the height of the right cylinder, the projection of the direction orthogonal to the plane a must be obtained; $T_{\alpha z}'$-1 is the projection of the line of maximum slope of α on xy, and ($T_{\alpha z}$)-1 is its overturning.

O-(2), conducted through O to ($T_{\alpha z}$)-1 is the reversal of the orthogonal for O to α; (2)-2, parallel to ($T_{\alpha z}$)-$T_{\alpha z}'$, identifies 2, and O-2 is the projection of the direction sought; note that O-(2) shortens into O-2.

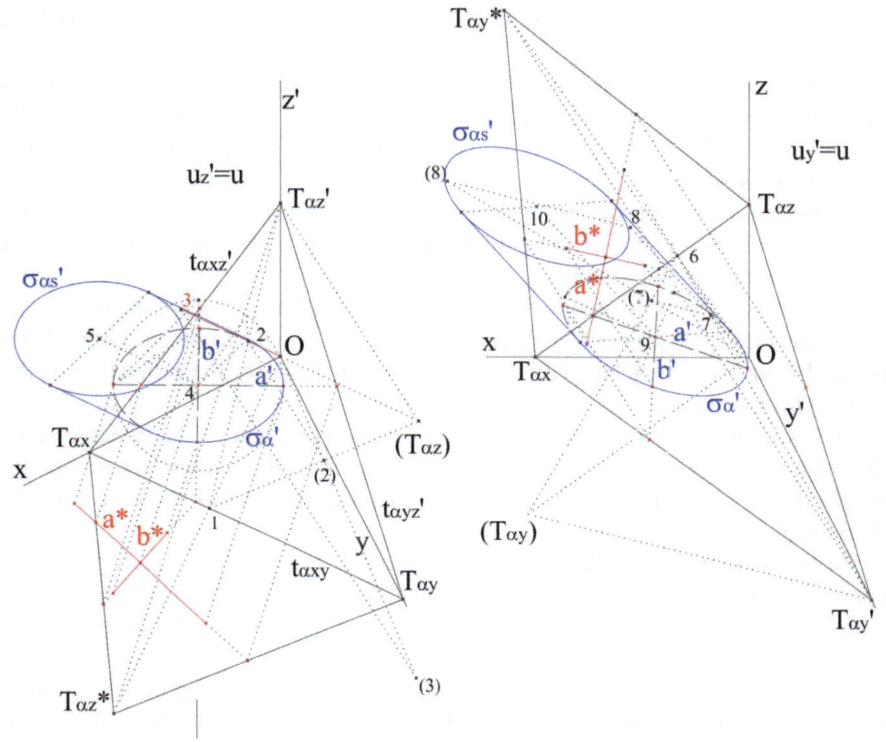

Fig. 4.11 Representation of a cylinder with $u_z' = u$ (on the left side) and $u_y' = u$

If the true dimension of the height of the cylinder is placed in O-(3), the relative projection is obtained in O-3, by similarity; O-3 is translated into 4-5; similarly, the ellipse $\sigma_{\alpha s}'$ of center 5 is obtained from the σ_α' of center 4.

Using homology, we obtain the projections of the apparent contour generators, whose lower extremities delimit the visible part of the base ellipse on α; the image of the cylinder is complete.

In the hypothesis of α parallel or coinciding with $\pi = xy$, the bases are in true form and magnitude, and the height parallel to z′, being $u_z' = u$, is in true magnitude; it is left to the scholar to choose different data and to deepen the investigation; for example, when does the position of α produce the projection of the greatest height of reality?

Another hypothesis: consider the reference triplet x, y′, z, O, with $u_y' = u$ and bring the distances of the tracks unchanged from O, obtaining $T_{\alpha x}$, $T_{\alpha y}'$, $T_{\alpha z}$; plane α preserves the position and the distance from O.

The $T_{\alpha x}$-$T_{\alpha y}'$-$T_{\alpha z}$ triangle looks different; its $T_{\alpha x}$-$T_{\alpha y}$*-$T_{\alpha z}$ flip to $\pi = xz$ by rotation around $T_{\alpha x}$-$T_{\alpha z}$ is the same as in the previous case but rotated.

Precisely, $T_{\alpha x}$-$T_{\alpha z}$ corresponds and is equal to $T_{\alpha x}$-$T_{\alpha z}$*, $T_{\alpha x}$-$T_{\alpha y}$* to $T_{\alpha x}$-$T_{\alpha y}$, and $T_{\alpha y}$*-$T_{\alpha z}$ to $T_{\alpha y}$-$T_{\alpha z}$*. These determine O-6, which is orthogonal to $T_{\alpha x}$-$T_{\alpha z}$, O-($T_{\alpha y}$)

$= O\text{-}T_{\alpha y}'$ which is orthogonal to O-6; $(T_{\alpha y})$-6 which is the true dimension of $6\text{-}T_{\alpha y}'$; 6 $(T_{\alpha y})$ is flipped into $6\text{-}T_{\alpha y}{}^*$.

With reference to the points of intersection of the axes a*, b*, with the sides of the triangle overturned, the aforementioned points are reported from the triangle $T_{\alpha x}\text{-}T_{\alpha y}\text{-}T_{\alpha z}{}^*$ to the $T_{\alpha x}\text{-}T_{\alpha y}{}^*\text{-}T_{\alpha z}$; the pair of given axes is reproduced in the same position and size.

As before, we obtain the conjugate diameters a′, b′, projection of the axes a*, b*, and, by obtaining the axes, we construct the ellipse σ_α', only apparently different from that of the previous case.

Consider O-(7) orthogonal to $6\text{-}(T_{\alpha y})$, (7)-7 parallel to $(T_{\alpha y})\text{-}T_{\alpha y}'$, the projection O-7 of the orthogonal for O to α; O-(7) shortens to O-7; in the same proportion from the real value of the unchanged height of the cylinder O-(8) the projection of the height O-8 is obtained.

O-8 is translated into 9-10 and σ_α' into $\sigma_{\alpha s}'$, a projection of the upper base of the cylinder; by homology, the projections of the apparent contour generators are obtained and from their lower extremities, the exposed part of the ellipse σ_α' is delimited.

Point 5, or point 10, can be the projection of the vertex of a right, square cone with parabolic points, having the elliptical base on α; the cone can be turned upside down, with the vertex on α; these are, however, cases that commit to simple reflections, close to banality.

Quadric conical surfaces must have a conic as a plane direction; consider (Fig. 4.12) a cone, whose base is a hyperbola, which is to be obtained from a circle by homology.

Let the axis and the limit line of the circumference be given, whose center is also the center of homology; Once the point 1 of the diameter 1-2 has been chosen on the circumference and a line by 1 has been drawn, let 3 and 4 be its points on the limit line and on the axis.

The corresponding homological of 1-4 must be parallel to 3-C; the corresponding V_1 of 1 must be on the aforementioned parallel and on the 1-C; to obtain the homologue V_2 of 2, the limit point 3 can still be used; the 3-5-2 is obtained and the parallel to the 3-C is led through the 5, which identifies the V_2 on the 2-C.

Consider the limit point 6; the homologation of a 7-6-9-8 string contains the joined point 9 and is parallel to 6-C; on this parallel the 7-C and 8-C identify the homologues 7′ and 8′; having chosen another string for the 6, the homologous, parallel to the 6-C for the joined point of the rope, identifies the homologues of the ends of the rope, joining them with C.

The homologues of the points of the circular arc 10-1-11 form the branch of vertex hyperbola V_1 and those of 10-2-11 the branch of vertex V_2, which intersects the circumference at the joined points common to it and to the axis.

The parallels at 10-C and C-11 for the mean 12 of $V_1\text{-}V_2$, the focal axis of symmetry, are the asymptotes of the i_{xy}' hyperbola; in the reference x, y, z′, O, with $u_z' = u$, it is the base of the right cone of vertex V and height C-V′.

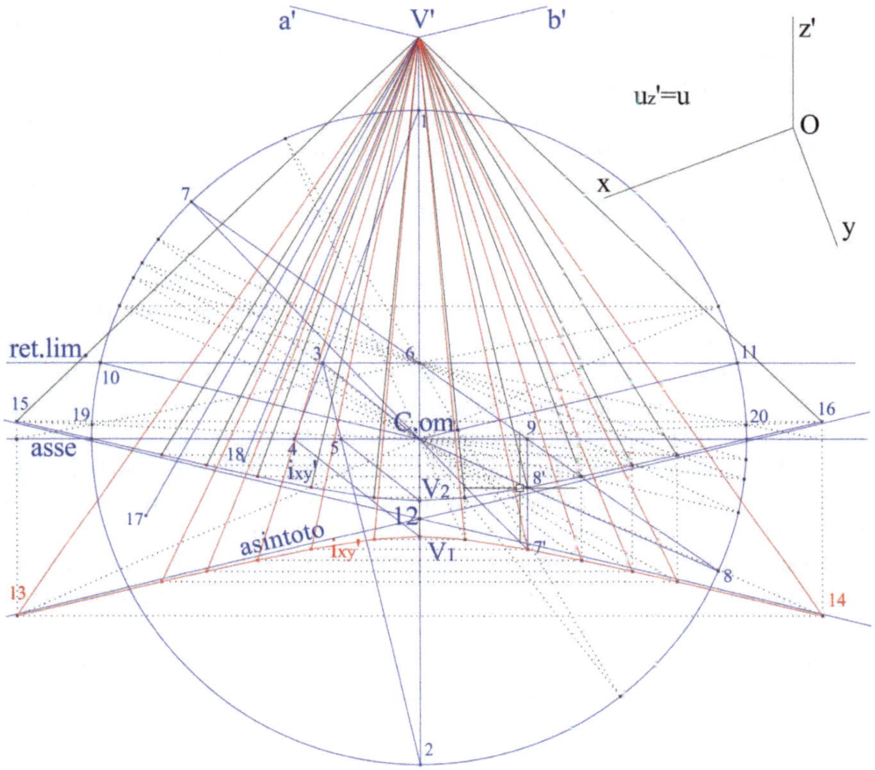

Fig. 4.12 Quadric conical surfaces whose base is a hyperbola, which is to be obtained from a circle by homology

Among the generators marked for the lower layer of the cone, the a′, b′, are those of apparent contour, parallel to the asymptotes; in the representation of both pitches, one would feel the overlapping of the relative images.

Reflect: V′-17 is outside the cone; the V′-18 is internal; to appreciate the convexity of the cone, one must consider the surface of a portion of the solid, for example, the base 15-V_2-16-15, corresponding to the circular arc 19-2-20.

The geometric surface, without thickness, has two sides; the solid side of the solid is theoretically given by the previous choice; but the opposite can be imagined; sections 15-13-V′ and 16-14-V′, closing the curved and conical surfaces 13-V_1-14 and 15-V2-16, delimit a visually identifiable volume, a truncated concave cone.

The hypothesis stimulates the imagination and stimulates lucubration, confirming the growing relationship between inspiration and reflection; In light of this, the example treated appears susceptible to some complications.

For example, the base may belong to a generic plane, or the reference does not contemplate xy as π, or, the trunk may perceptually be considered derived from a pyramid with a rectangular base 13-14-16-15-13, in which a cone has been dug on a pair of opposite sides having an arc of hyperbola as its base; etc.

Among the infinite possible variations, a suggestion, a request, the insinuation of an "if" ignite curiosity and refer to commitments of clarification, in-depth analysis, conditioning, learning, orientation to the desire to know.

Example is not, cannot and must not be an end in itself; it is like the crumb thrown into the sea, triggering a frantic rush of fish, each respecting the biological chain to which it belongs.

The scholar's attention should not escape the continuous effort to achieve the objective of this treatise, which, at the same time, is based on active preparatory knowledge, promotes access to higher-level knowledge.

It is necessary to take advantage of references, exhumations of concepts and principles, general and particular, of the basic elementary platform of the geometry of drawing, really acquired and experimented in the degree course.

It is also necessary to devote oneself to careful reworking of one's own baggage of scientific and psychic availability, restructuring them, consolidating them, expanding their relationships and functions, never in a definitive form.

Consider (Fig. 4.13), referring to x′, y, z, O, with $u_x{}' = u$, a right cone of vertex V_1, having on xy the circular base b_1 of center C_1 and radius r_1; use the oblique affine homology defined by the pair of diameters parallel to x′ and y; the minor axis and the semi-major axis are obtained and the elliptical image of the base is obtained.

Once the height has been reported and the apparent outline generators have been obtained by homology, one can foresee the embarrassment of the layman in knowing how to read the figure correctly; the sight certainly does not remind him of a straight circular cone, of common memory!

Considering a rectilinear radius for the vertex, let the tangents at the base $b_1{}'$ be conducted through the intersection $V_1{}^*$ on xy; The generators for the contacts are shadow separators.

We want to underline the problem, often underestimated or misunderstood, of the correct reading of images; consider the right cone, vertex V_2 of a given height, and whose base b_2 on xy appears to be a circumference.

The usual layman would instantly arrive at his obvious reading of looking at a straight circular cone; We must be very careful, it is not enough to look, but it is necessary to evaluate, reflect, think, reason, doubt, verify, and discover.

By reversing the pair of axes having the direction of the diagonals of the rhombus, an image of the square resting on the x, y axes, the homologous ones are obtained; the real base is elliptical; it is, therefore, not a right circular cone, but an elliptical cone.

Then you can teach how to say "it seems to me"; It is significant of wisdom to try to identify clues that exclude that assessment of fallacious impetus and to say, instead, "it would seem…., but it is…".

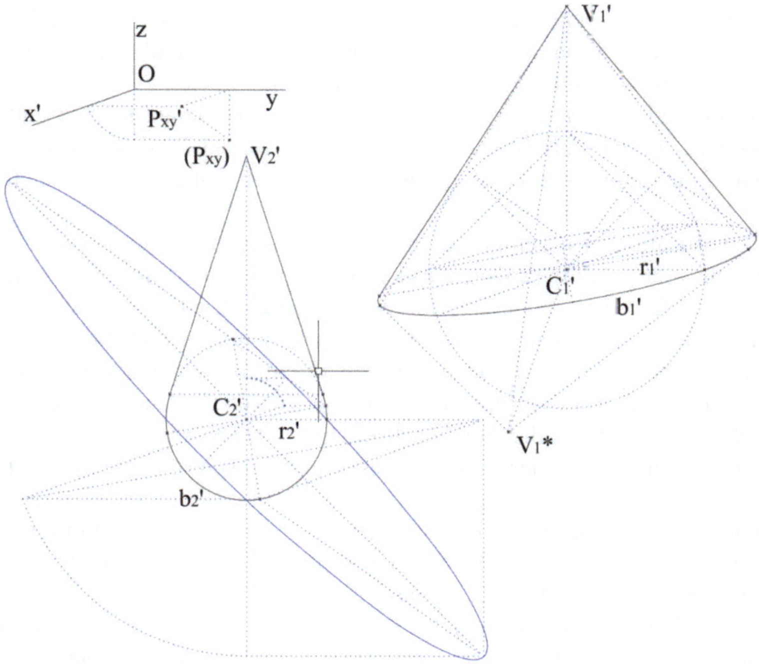

Fig. 4.13 Representation of a right cone of vertex V_1, having on xy the circular base b_1 of center C_1 and radius r_1, with $u_x' = u$, and of a right cone of vertex V_2 having on xy the elliptic base b_2 of center C_2 and radius r_2, with $u_x' = u$

Elliptical Point Quadrics

If we want to represent a quadric with elliptical points in oblique axonometry: we instinctively think of the sphere, and excluding this, the ellipsoid or, rarely, the elliptical paraboloid or the elliptical hyperboloid.

Consider a brief definition of the possible spatial genesis, correctly chosen for representative purposes; for the sphere, for example, it can be said that for each of its points, the distance from a center point of the sphere is constant.

Another definition of the spatial genesis is given by the rotation, for 180°, of a circumference, chosen as the maximum generating section, around one of its diameters; or again, the equation of the spherical surface can be assigned.

In the latter case, having or giving the idea of the form is euphemistic; In the first statement, a star of rays of equal length, by obvious necessity, in discrete quantities, even if it corresponds to a balanced spatial distribution, cannot satisfy the imagination of the spherical shape.

Finally, the rotational genesis of the maximum circular section around its diameter appears less enigmatic, but, in any case, it is not entirely satisfying result of an analytical and reflective effort.

If, then, a representative question is proposed, for example, how the image of the sphere will appear on the painting, one remains dumbfounded, recognizing the simplistic and hasty validity of the a priori definition.

It is believed that the reference to the sphere, a redundant surface and space of particular and particularly invariant qualities, is already a simplification of the problem of the representation in orthogonal axonometry of quadrics with elliptical points.

The clarification “with elliptic points” declares that the solution of the system of equations relating to the square surface and a tangent plane results in a negative discriminant; The lines common to the two aforementioned geometric entities are imaginary and have in common, and real, only one point, the real one of tangency.

It can also be said that with respect to the aforementioned tangent plane all the points of the quadric are all positioned on one side of the tangent plane; This is essential knowledge to predict how much and how it is necessary to reflect, to know and, again, to reflect, in order to orient oneself to the result of the representation on the subject.

It would be necessary to deal with the entire group of elliptical point quadrics, examining their peculiarities in full and comparatively; reasons of expediency advise limiting oneself to the study of the sphere.

Consider (Fig. 4.14) the axonometric reference x, y, z′, O, with $u_z' = u_z$; the center C of the sphere is identified by the projections C′, C_{xy}'; if it coincides with O, its projections are C, C_{xy}.

The polar axis N-C-S belongs to z and projects into N′-O-S′, and being $u_z' = u_z$, it is in true magnitude; consider the plane γ projecting z into z′, of traces $t_{\gamma xy}$, z′.

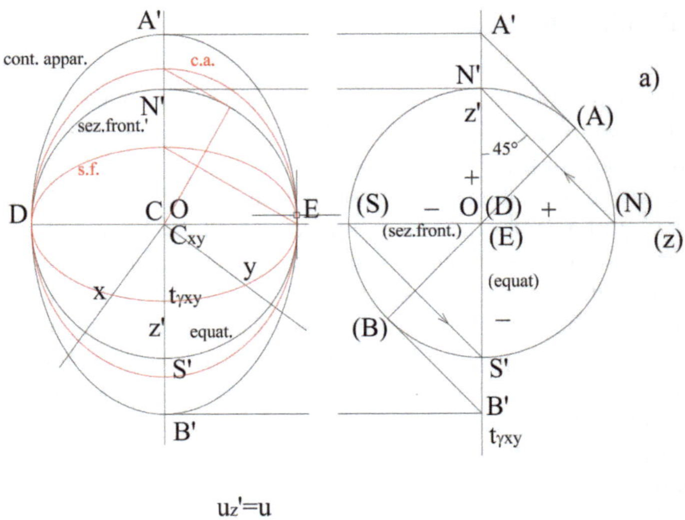

Fig. 4.14 Representation of a sphere in different view

Separately, on the right, flip γ over xy; (z) is orthogonal to $t_{\gamma xy}$, and the inverted polar diameter is given by (S)-O-(N); the ray that projects (N) at 45° on xy obtains N′ on $t_{\gamma xy}$, with O-N′ = O-(N); similarly (S) projects into S′.

In the maximum circular section of the sphere, belonging to the inverted plane γ, from the diameter (A)-O-(B), orthogonal to the direction of projection (N)-N′, we obtain A′-O-B′.

Each of the infinite diameters of the aforementioned section, having the projection on $t_{\gamma xy}$, corresponds to a projection, varying from a minimum value of zero, relative to the diameter for O parallel to (N)-N′, to the maximum value; this is not relative to the polar axis, but to (A)-(B).

The diameter (D)-O-(E) of the sphere, orthogonal to the plane γ, is a point in the overturning, while, in projection, it is a frontal diameter D-O-E, in true magnitude.

The circular section defined by the two orthogonal diameters (A)-O-(B) and (D)-O-(E) projected on xy is an ellipse, of which A′-O-B′ is the major axis and D-O-E is the minor one.

The infinite projection radii, such as (N)-N′, tangent to the sphere, constitute the tangent cylinder projecting the sphere onto xy and identify the contact circumference defined by the diameters (A)-O-(B) and (D)-O-(E); The aforementioned cylinder intersects the xy frame not according to a circumference but according to an ellipse.

All points of the sphere have the projection inside the aforementioned ellipse, which, therefore, is called the "apparent contour" of the projection of the sphere.

The equatorial section is projected according to an "equat" circumference of center O and true radius; in the case represented, it coincides with its projection; The polar frontal section is projected according to a "front" circle and overlaps the previous "equat".

It can be observed that the semi-major axis O-A′ is equal to the diagonal of the square of the radius O-(A); the highest point of the sphere, in projection is N′ and does not appear as such; the point A′ represents a point A of altitude lower than N.

The "equat" projection of the equatorial circumference is always a circle, belonging to xy; the polar axis is in true magnitude only if you choose to project at 45°; if the projection radius conducted for (N) on z′ forms an angle greater than 45° with it, it is O-N′ < O-(N) and the frontal polar circumference projects into an ellipse, defined by the major axis D-O-E and the minor N′-O-S′.

Conversely, if the angle is less than 45°; reflect on the different position assumed by A′, which is obtained by leading the orthogonal to the different inclination of (N)-N′ by O; obtained the different (A), the parallel a (N)-N′ per (A) identifies on z′ the different A′; for angular value tending to 90°, N′ tends to O and A′ to N′.

In Fig. 4.15, the projection of the spherical surface, with diameter D-C-E, is proposed again when the center C(C_{xy}, C_{xz}', C_{yz}', C_x, C_y, C_z') of the sphere is not chosen in O.

The equatorial circumference, equat., of the sphere, of center C, is parallel to the square xy; to be identified spatially, the circular projection, equat_{xy}, orthogonal to the xy frame, with center C_{xy} and diameter equal to D-C-E, is also considered.

Fig. 4.15 Sphere with vertical axis: True-size projection of equator when polar axis is vertical

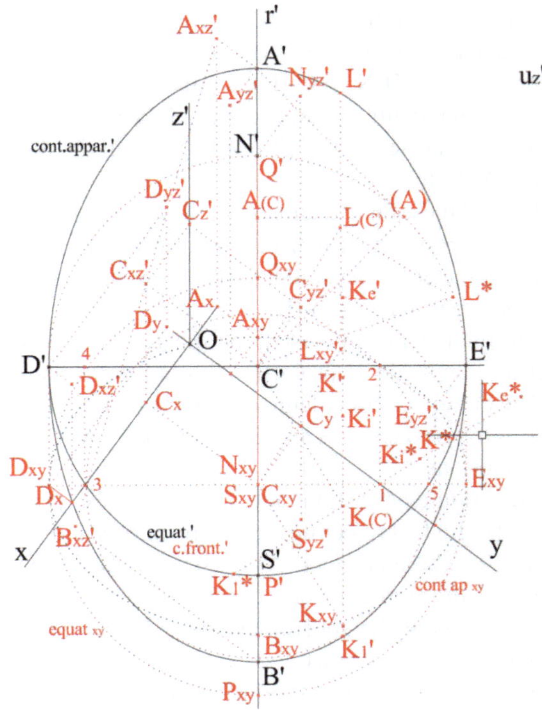

The oblique projection at 45° on the xy frame of the equatorial circumference is the equated circumference, of center C′ and diameter D′-C′-E′; that of its circular projection of the center C_{xy}, belonging to the painting, coincides.

Reflect on the projections of the diameter of the equatorial circumference, orthogonal to D-C-E; the pairs of projections, P′, P_{xy}, and Q′, Q_{xy}, represent the extremes; P is the point of the sphere closest to the observer, and Q, the farthest point.

Consider the frontal circumference, c_{front}, parallel to z, defined by the diameter D-C-E and the vertical diameter, polar axis N-C-S; the first is parallel to the xy frame and is projected into D′-C′-E′, in true size, the second, being parallel to z and for u_z' = u_z, is also projected into true size, into N′-C′-S′.

The front is circular and superimposed on the equated circumference; reflect on the substantial difference of the two diameters P′-C′-Q′ and N′-C′-S′; N_{xy}-C_{xy}-S_{xy}, is the orthogonal projection of N-C-S, and its axonometric projection is a point.

Note that with respect to the xy table, the extreme N has the positive dimension N_{xy}-N′ and S has the negative dimension S_{xy}-S′; reflect on the fact that the sphere is intersected by xy according to the smaller circumference c_{xy} of center C_{xy} and radius C_{xy}-5, of a value lower than that of the sphere C′-E′.

If the dimension C_{xy}-C′ becomes equal to C′-E′, i.e. S′ is found in S_{xy}, the sphere is resting on the xy frame; it is noticed, however, that S′ image of the lowest point of the sphere will never be the lowest point of the image of the sphere, which is, instead, B′, the lower end of the major axis of the ellipse apparent contour of the sphere.

The same is true for N′, an image of the highest point of the sphere but not the highest of the image of the sphere, which is, instead, the point A′, the extreme of the major axis considered above.

Reflect and not share, without reflection and reasoning, the conclusions of considering the present projective process as the simplest and most accessible, with respect to perspective, orthogonal axonometry, dimensioned projections, and multiple orthogonal projections.

Pay attention to appearances, do not mistake what appears, or what seems, with what is; reality does not arise from appearance; To conclude early, erring, after warnings, may be inadmissible arrogance.

Identify on r′, parallel to z for C′, the point A′ by means of the tangent in (A), obtained from C′-(A) orthogonal to the reversal E′-N′ of the direction of the projection on the frame, respecting the projective choice adopted $u_z' = u$

The point $A_{(C)}$ is at the level of C′ and $A_{(C)}$-A′ is the value to be added to the value equal to C_{xy}-C′ in order to obtain in A′-A_{xy} the share of A with respect to xy; A_{xz}' and A_{yz}' derive from this; the three coordinates on the three coordinate planes of A are known and those on the axes are obtained.

The y-parallel to B′ intersects A_{xz}'-C_{xz}' in B_{xz}', symmetric of A_{xz}' with respect to C_{xz}'; we can proceed from A_{xy} by obtaining B_{xy}, symmetrical with respect to C_{xy}; By working carefully, the extremes of the projections of the diameters and the coordinates of each of their points can be obtained.

From the projections of the points it is possible to verify that they belong to the spherical surface, since their distance from the center must be equal to the radius; for example, considered, orthogonally on xy, the elliptical projection "cont ap $_{xy}$" of the apparent circular outline of the sphere, which in projection on the axonometric frame is the ellipse "cont.appar.", of axes A′-C′-B′ and D′-C′-E′, the relative axonometric projection is given by the ellipse "cont ap $_{xy}$" of axes A_{xy}-C_{xy}-B_{xy} and D_{xy}-C_{xy}-E_{xy}.

Chosen L on the apparent outline ellipse "cont.app.′", follows L_{xy}' on the "cont ap $_{xy}$"; the dimension of L on xy is L_{xy}'-L′ and at the level of C it is $L_{(C)}$-L′; the latter being orthogonally to C′-$L_{(C)}$ for $L_{(C)}$, we obtain in C′-L* the distance C-L equal to the radius; in fact, L* falls on the "equat" circumference of radius C′-E′.

Once a point K(K′, K_{xy}) has been assigned, we want to verify whether it belongs to the surface of the sphere; it is outside it if the projection K′ is outside the "cont appears", if $K_{(C)}$ is outside the "equat′" circumference, if $K_{(C)}$-K′ is greater than the radius C′-E′, if C_{xy}-K_{xy} is greater than C′-E′.

Let it be reflected; these checks, sometimes even by eye, can lead to an immediate response of exclusion; not putting it before the verification operations is proof of unedifying superficiality.

Let K_{xy} not fall within any previous case of exclusion and give a value to the dimension of K so that K′ represents a point on the spherical surface, i.e., an internal or external point.

From C_{xy}-K_{xy} we arrive at C′-$K_{(C)}$ through K_{xy}-$K_{(C)}$ = Cxy-C′; the line orthogonal to C′-$K_{(C)}$ for $K_{(C)}$ intersects the "equat" circle in K*; $K_{(C)}$-K* is returned to $K_{(C)}$-K′, locating in K′ a point on the spherical surface of coordinate K_{xy} on xy and dimension K_{xy}-K′.

Of the point K, let the generic dimension K_{xy}-K_i' or K_{xy}-K_e' be given on K_{xy}; at the end of the verification, the first is internal and the second external: observe that $K_{(C)}$ is the average of the chord K^*-$K_1{}^*$; $K_1{}^*$ corresponds to K_1', whose altitude is K_{xy}-K_1', negative.

Reflect; of the vertical line for K_{xy} the points of intersection with the sphere can be obtained; the points K^* and $K_1{}^*$ and their straightening in K' and K_1' are obtained by overturning for $K_{(C)}$, orthogonal to C'-$K_{(C)}$.

It should be noted that K_1' appears to belong to the ellipse "cont.appar′"; The apparent outline of the sphere image is the apparent outline image of the sphere.

The representation of the sphere was proposed, imagining the polar axis parallel to the z axis; nothing prevents the choice of another direction; if in the vertical projecting circular section, overturned in place of (A) we think (N), for $u_z' = u$, N′, and therefore S′, are on the apparent boundary of the projection of the sphere.

They are extremes of the major axis of the ellipse, which also represents the frontal polar section forming with z the angle of 45° adopted for the oblique axonometric projection.

In the case discussed here, the projection of the equatorial circumference is the diameter D′-C′-E′; on the sphere, there is the circular section of center C and parallel to the xy frame, whose projection is preserved circular; considering it is irrelevant.

In Fig. 4.16, an axonometric projection direction is placed on the xy frame no longer at 45° but of a greater value φ, so that its apparent contour on the sphere is a trace plane D-C-E; its trace on the projecting plane, vertical for C and orthogonal to D-C-E.

The circular section of center C, belonging to this plane, is overturned, and on it (A) is identified, such that the C′-(A) and the tangent in (A) verify that the angle φ remains defined in A′.

The point A′ and the symmetric B′ with respect to C′ are extremes of the major axis of the apparent boundary ellipse, D′-C′-E′ being the minor axis; the major axis depends on the value of φ.

Once the reference x, y, z′, O has been established and the unit value u is placed on x and y, this value u is plotted orthogonally to z′ and, projecting according to (A)-A′, u_z' is obtained; it follows that the polar axis shortens to N′-C′-S′, being E′-N′ parallel to (A)-A′.

In the sphere, resting on xy, S and C_{xy} overlap; followed by the other two coordinates C_{xz}' and C_{yz}' of center C; as long as the N-C-S polar axis is parallel to z, the equatorial section is parallel to xy and projects into true shape and size.

Note that the 2 is obtained on C′-N′, chosen C′-1, projecting it according to the 1-2, parallel to the E′-N′; it is the center of the circular projection par_1' of the minor section parallel to the equator, at a distance from C equal to C′-1, and of radius 1-3, all in sight; It can be said that the par_1' has no points of contact with the apparent boundary of the projection of the sphere, i.e. that it has two imaginary contacts with it.

The 5 is obtained on C′-N′, chosen C′-4, projecting it according to the 4-5, parallel to the E′-N′; it is the center of the circular projection par_2' of the minor section parallel

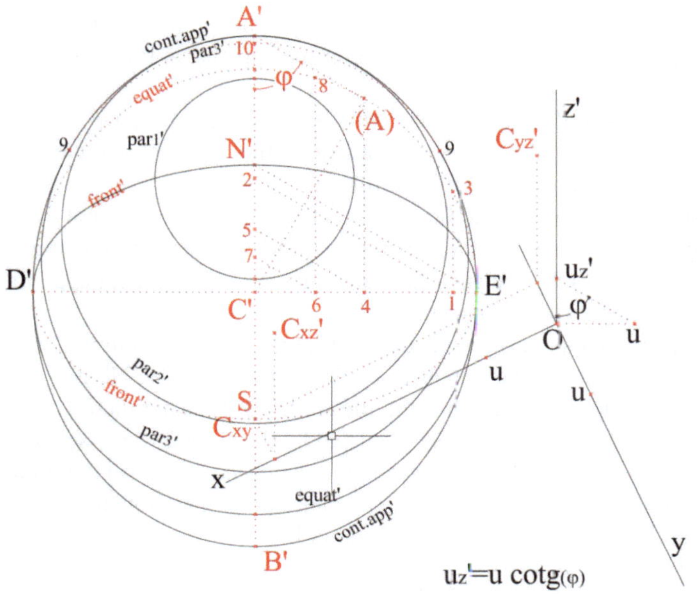

Fig. 4.16 Representation of the surface of the sphere where there is the prevalence of the visible part, linked to the pole

to the equator, at a distance from C equal to C′-4 dependent on (A), and of radius 4-(A), all in sight.

This circumference touches in (A) the apparent outline and is the largest parallel still in sight; It can be said that it is the first and only section with a double contact joined.

The 7 is obtained on C′-N′, chosen C′-6, less than C′-4, projecting it according to the 6-7, parallel to the E′-N′; it is the center of the circular projection par_3' of the minor section parallel to the equator, at a distance from C equal to C′-6 and of radius 1-3, in partial view, since the arc 9-10-9 is not in sight.

Symmetrically, the arch entirely in view corresponds to the arch entirely not in sight; the arch all in sight and with a point of contact with the apparent contour corresponds to the arch all hidden and with a point of contact; the major arc of the semi-circle, all in view and with two points of contact, corresponds in view to only one arc, smaller than the semi-circle, bounded by two points of contact on the cont.app′.

Of the surface of the sphere there is an evident prevalence of the visible part, linked to the pole that is visible; the equated circular projection of the equatorial section is half visible, provided that the polar axis is parallel to z′; For this reason alone, the projections of the equator and parallels, being parallel to the XY frame, retain shape and size.

Think of a decomposition of the sphere into portions: the delimitation of the parallel par_1' which, in the figure, includes N′ and a short arc of the frontal section front′, for N′, represents a cap of the sphere with the highest point in N′; there is no trace of the apparent outline of the sphere; the cap shows the par_1' as an apparent outline.

The boundary of the parallel par_2', which contains A′, N′, par_1', and an arc of the front. It is the spherical cap that touches in A′ the apparent contour of the sphere only in A′; there is no view of this outline.

The delimitation of the parallel par_3' excludes the arc 9-10-9 not in sight; correspondingly, the arc 9-A′- of the apparent outline of the sphere remains visible; the spherical cap shows N′, the par_1' the par_2' tangent in A′ the cont app′, the arc front for N′, which approximates its half.

The upper half sphere is bounded by half of the equatorial circumference in front of the center C of the sphere; half of the cont.app′ D′-A′-E′, N′, par_1', par_2', par_3', half D′-N′-E′ of the front′ are in sight; the arc D′-B′-E′ is not there.

Reflect on the possibility of being misled by the simple visual acknowledgment: there is the cap delimited by the apparent entire outline; the lower delimitation, given by the circumference, unmarked and symmetrical to par_2', touches the cont.app′ in B, but is entirely hidden, except for the point B′.

Reflect again on the possible truncation of the sphere with parallels dissecting the lower hemisphere and not visible; if there is no clue to their existence and delimiting function in the figure, it is not possible to read the actual volume.

The constructed graphic image, unlike photography, obviates this serious drawback, reporting in dotted, for lawful transparency, everything that has interpretative value, but is not, due to the exhibition position, significantly visible.

It is not a question of superiority of one over the other; Those who read without reason, without posing the problem of the adherence between appearance and reality, without being bound to the certainty of an irrefutable response of recognition, assume sufficient the messages of photography and not of the geometry of form.

The photographic image, in the end, offers very different messages, of irreplaceable and indispensable cognitive contribution, not reported and not reported in the geometrically constructed image.

The interest of considering the view of a spherical area, removing a cap from a cap, is not negligible; similarly, with the consequent adjustments, it can be argued on the spherical bowl, starting from the lower pole S′, on the spherical bowl, on the spherical wineskin.

And again, we can hypothesize the view from below of both the caps and the cups; the concave surface of the former is exposed and the convex surface of the latter; the flat section that delimits them can have a horizontal, vertical, or generic position.

The geometric problems on the sphere cannot fail to deepen the intersection between line and sphere, between plane and sphere, the distance between point and sphere, the tangent planes conducted through a line without real contacts, and the tangent cone, which has a given external point as its vertex.

Reflect on the ambiguity of the external position, remembering the properties of polarity; In general, the need to extend the investigations a little beyond the usual, the easy, the common, the compliant, escapes.

Sufficient problems involving sight, shadow, reflected image, movement, composition, decomposition, interpenetration, cavity, holes, symmetries, concentricity, concatenation, coaxiality, etc., should be addressed.

Consider (Fig. 4.17) the sphere of center O in the frame of reference x, y, z′, O; determine the projections of the apparent contour "cont.app′", of the equatorial circumference "equat", and of the frontal polar section "front′".

Assign the line r defined by the traces T_{rxy}, on xy, and T_{ryz}' on yz; We want to verify whether r has in common with the spherical surface two distinct real points, for which it is incident, two coincident real points, for which it is tangent, two distinct imaginary points, for which it is said to be external, improperly.

Let F and G be common to the "equat" and to the r_{xy}, which contains T_{rxy} and H on y, projection of T_{ryz} on xy; if the pair F, G, is imaginary, the line does not intersect the sphere.

The chord F-I-G is diameter of the vertical circular section c_r, coplanar with the r; the vertical beam projects into I-L′, conjugated to F-I-G; the ellipse c_r' is derived; it has contacts in 1 and 2 with the ellipse "cont.app′", not necessarily; in the case in question, the 1-3-2 arc is in sight; The rest is hidden.

The lack of contact between r′ and c_r' is observed; the line does not intersect the sphere; one can hypothesize the motion of r on its vertical projecting plane, rotating

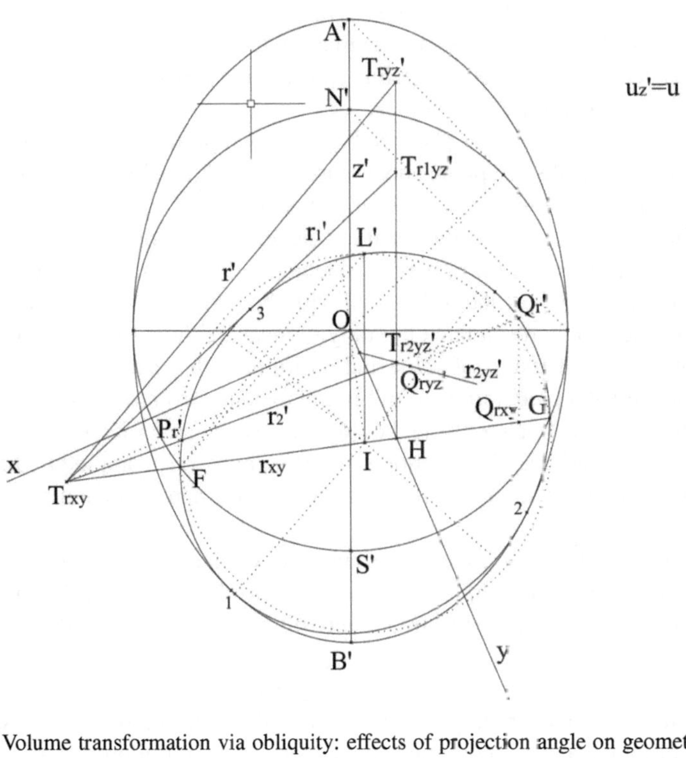

Fig. 4.17 Volume transformation via obliquity: effects of projection angle on geometric volume appearance

it around T_{rxy}; at the position r_1', the line touches the sphere in 3 and its trace is T_{r1yz}'; at the position r_2', the line intersects the c_r' in P_r' and Q_r', distinct points common to the sphere.

Observe the derivation of r_{2yz}'; Q_{ryz}' belongs to r_{2yz}' and Q_{rxy} to r_{xy}; imagining and posing different hypotheses serves to evaluate whether one possesses, and at what level, full consciousness of science.

Given two points K, J, on the line r, in the case of a line outside the sphere from K is seen J and vice versa; in the case of the tangent line, we can still say that J is seen from K, or directly, if both are positioned on the same side with respect to point 3.

In the case of an incident line in P and Q, in the sequence K, J, P or K, J, Q, from K is seen J, in the sequence K, P, Q, J from K is not seen J; The problem arises in the question of whether you see if the points are given or you don't want them to be seen.

Of the line, the stretch P-Q is inside the sphere and must be dotted; the stretch from the apparent boundary to the intersection P is outside the sphere; the entire line must be used up to P; from Q to the apparent boundary is external but hidden; it is highlighted with a line in sections.

Regarding the use of lines, note that the arc D-N′-E is a full-length if it belongs to the "front" and the remaining arc is a full-length if it belongs to the "equat"; reflect on the easy incompetence of reading.

In the figure, the mandatory alignment with the correct use of the lines is not always respected; indeed, in some points of the final image, it is admittedly disregarded, to help the reader in the improbable effort of orienting himself among the countless profuse or implied graphic testimonies.

Knowing how to derive the distance between two points, the minimum distance between a point outside the sphere and its surface, we obtain, by subtracting the value of the radius from the one between point and center.

Knowing how to lead the orthogonal to a generic plane for a point, the minimum distance between this and the sphere is reduced to the previous one, first obtaining the point of intersection of the orthogonal with the assigned generic plane.

The exercise can be formulated not only to measure, but also to derive the initial data to obtain desired results of value and position; in the case of the generic plane, it can be desired at a predetermined minimum distance from the sphere.

In the figure of the line r, outside the sphere, we can ask for its minimum distance from the convex surface; The solution process is, however, completely different; in fact, this value must be sought on the plane for the line and the center and not on the vertical projecting plane.

The section is, therefore, the maximum one, coplanar with the straight line; a reversal or a rotation and a lot of reflection lead to the resolution; In any case and to avoid misunderstandings, the treatment follows, which can and will never be exhaustive.

In Fig. 4.18, in the reference frame x, y, z′, O, $u_z' = u$, there is a sphere of center O and radius O-D, and three points (not marked in the figure); of the sphere we obtain the elliptical apparent outline "cont appar′", the circumference on xy "equat",

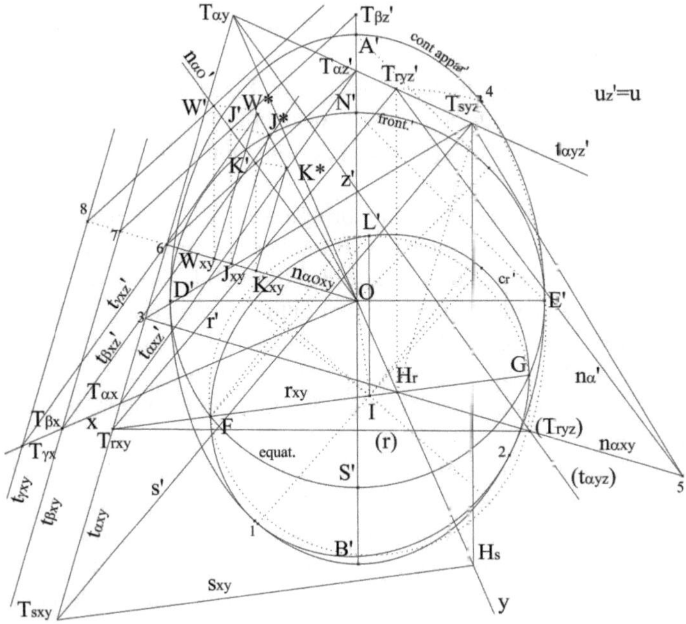

Fig. 4.18 Representation of sphere with the elliptical apparent outline

in view only for half D′-S′-E′, the circumference for z "front.", only half of D′-N′-E′ is in sight.

The three points belong to a plane α; for two of them the line is considered and for the third the parallel to it; thus, the line r (r', T_{rxy}, T_{ryz}', H_r, r_{xy}), the line s (s', T_{sxy}, T_{syz}', H_s, s_{xy}) are defined; from α which we obtain $t_{\alpha xy}$, $T_{\alpha x}$, $t_{\alpha yz}'$, $T_{\alpha y}$, $t_{\alpha xz}'$, $T_{\alpha z}'$.

We want to check if α cuts the sphere, to derive the possible distance from the sphere or from its center, possibly translate the plane α, until it is reduced to a tangent plane β or to a plane γ, at a distance assigned from the sphere or its center.

Out of curiosity, the construction of the position of r with respect to the sphere is reported, determining its vertical section, circular c_r in space, but elliptical c_r' in projection.

It should be noted that the verification according to which r has no real points in common with c_r has a meaning, in truth reduced, only if r were to be incident; in this case, the incidence between α and the sphere would certainly result; the plane would have a distance from the center of the sphere less than the radius.

Since r is external to c_r, this is not sufficient to conclude a correct answer; in fact, the plane of c_r should contain the center of the sphere; if the incidence between r and the section between this plane and the sphere were to result, it could be said, in this case, that the distance between α and the sphere is less than the radius.

It should also be noted that the r outside this section would certainly not imply the intersection between α and the sphere; it is correct to proceed, considering the

n_α ($n_{\alpha O}'$, $n_{\alpha Oxy}$) orthogonal to α for O, obtaining the point K of intersection and comparing O-K with O-D.

The $n_{\alpha Oxy}$, orthogonal to the $t_{\alpha xy}$ in 6 and the $T_{\alpha z}'$-6 define in projection the line of maximum slope, on α and for $T_{\alpha z}'$, with respect to xy; the orthogonal to the plane is defined by the directions $n_{\alpha Oxy}$ and $n_{\alpha O}'$, orthogonal to $T_{\alpha z}'$-6; it intersects α in K (K′, K_{xy}).

The true distance O-K is obtained by bringing K′-K_{xy} back into K_{xy}-K*; it is less than O-J* = O-D; the plane α intersects the sphere according to a circumference of center K* and diameter equal to the chord led through K* to O-K*; its projection is an ellipse of center K′ and major axis equal to the aforementioned chord.

If we want the plane β, parallel to α and tangent, we obtain J*, J_{xy}, J′, J′-7-$T_{\beta z}'$ parallel to the 6-K′-$T_{\alpha z}'$, the 7-$T_{\beta x} = t_{\beta xy}$ parallel to the 6-$T_{\alpha x} = t_{\alpha xy}$, for $T_{\beta z}'$ the $t_{\beta yz}'$ parallel to the $t_{\alpha yz}'$, not marked in the figure; the plane β, tangent to the sphere in J is defined as follows.

If we want the plane γ of position α and at a distance from O equal to O-W greater than the radius of J*-W*, we obtain W_{xy}, W′, W′-8, 8-$T_{\gamma x} = t_{\gamma xy}$, $t_{\gamma xz}'$; the plane γ thus defined has no real point in common with the sphere and has the minimum distance from the spherical surface equal to J*-W*, in the projection J′-W′.

The burden of obtaining the section between α and the sphere is left to the scholar; the relative ellipse, depending on the distance from O, can result, with respect to the apparent contour, without real common points, or with a single contact, or with two real points, which divide the relative ellipse into two arcs, one in view and the other hidden.

A variant to the question may consist in wanting to construct the plane for the line r and resting on the sphere; for r, without real points common to the sphere, two tangent planes can be conducted, symmetrical with respect to the plane for the line r and the center O.

The plane for O is orthogonal to the line r, intersects it at a point, and dissects the sphere according to a maximum circumference, at which, from the point trace of r, the two tangents are considered; The two planes for R and for the two tangents are the required tangent planes.

The exercise is conceptually affordable, using the usual inevitable attention and reflection; Most likely, however, the traces and points to be available in the resolution process are out of the field, causing unpredictable complications.

Appendix
Research Future Directions

Topics of Oblique Axonometry, the future treatment of which, possibly not carried out by the author, is reported and entrusted to the scholars.

The ellipsoid–The elliptical paraboloid–The elliptical hyperboloid–The hyperbolic paraboloid–The hyperbolic hyperboloid–Fans–Helicoids–Tori–Threads.

Flat and humpback, pleated and corrugated, rotational, rototranslation, pavilion, cross, dome, star, simple and multilayered lattices, flat and polyhedral, mushroom, tent, tensile structures, conjugate; Moebius strip.

Twisted column–Serpentine–Screw of St. Gilles–Cavities–Overhangs–Chamfers–Fittings–Pendentives–Cornices–Moldings–High reliefs–Bas-reliefs.

Ramps–Junctions–Stairs–Tunnels–Tunnels–Door and window openings–Arches–Lintels–Bridges–Piers–Scaffolding–Pylons–Viaducts–Overpasses–Underpasses.

Fencing, foundation, containment, consolidation, damming, adduction, disposal, parking, storage, storage, monitoring, moving, lifting structures.

Structures for health, sport, tourism, justice, administrative, industry, commerce, defence, energy, transport, entertainment, education, protection of the soil, coasts, the environment, etc.

Orographic, cartographic, cadastral, geological, archaeological, and statistical representations.

If you want to have patience and tenacity: everything that has been dealt with in **Perspective** and **in Orthogonal Axonometry**, it is the exhibitor's idea to deal with it also, and with increasingly lucid experience in **Oblique Axonometry**, in **Dimensioned Projections**, in **Multiple Orthogonal** Projections (of the Monge).

And that's not all; the integration programmed above starting from the ellipsoid is to be treated, in all the projection systems mentioned above.

And that is not all; analogous exposition, in all projection systems, of the **theory of shadows** and the **theory of reflected images**, profoundly innovated and therefore differently indicated, developed, and used, cannot be omitted.

L. Inzerillo, *The New Horizons of Oblique Axonometry*, UNIPA Springer Series,
https://doi.org/10.1007/978-3-032-07475-1

In fact, it is preferred not to speak of shadows but of incident energy, emitted according to beams of rays from its own or improper, point or multiple, fixed or mobile source, in order to study, research, prevent or verify its effects, or those effects, on large or small, fixed or mobile receiving surfaces, etc.; Energy can be light, acoustic, thermal, wind, water, mechanical, radio electronic, laser, biological, psychological, etc., adequately applied to specific problems.

In the same way, it is preferred to investigate geometrically the aforementioned energies emitted by any source and reflected by any element in any positional, dimensional, temporal situation, in the presence of any receiving subject, able to grasp their existence or to be prevented from doing so.

The author has already developed considerable innovations in his long teaching activity, which, in further studies, considerations, and reflections can mature into knowledge to be disseminated, promoted, indicated, stimulated, involved, interested, and excited.

Bibliography

Albeggiani M (1926) Applicazioni della geometria descrittiva, Palermo. Quaderni di Dipartimento

Bonanno L, D'Alessandro M (1985) Fra geometria e percezione, Facoltà di Ingegneria, Palermo. Quaderni di Dipartimento

Botvinovska S, Zolotova A, Mostovenko A, Sulimenko H (2022) Construction of hyperbolic paraboloids according to a prospective outline in the form of hyperbola, pp 207–216

Brion-Guerry L (1962) Jean plerin viator sa place dans l'histoire de la perspective, Paris

Capitano V (1972) Applicazioni di geometria proiettiva e descrittiva nel disegno delle forme geometriche elementari, Palermo

Capitano V (1996) Considerazioni omologiche nella prospettiva centrale su quadro inclinato, Facoltà di Ingegneria di Palermo

Csima G, Szirmai J (2016) Isoptic surfaces of polyhedra. Comput Aided Geom Des 47(55–60):2016. https://doi.org/10.1016/j.cagd.2016.03.001

Cundari C (1983) Teoria della Rappresentazione dello spazio architettonico, applicazioni di geometria descrittiva, Roma

D'Alessandro M (1979) Omologia nella prospettiva della sfera, Collana di studi di disegno della Facoltà di Ingegneria di Palermo. Quaderni di Dipartimento

Docci M, Migliari R (1992) Scienza della rappresentazione, Roma

Docci M, Gaiani M, Maestri D (2011) Scienza del Disegno. Città studi edizioni, Novara

Docci M, Gaiani M, Maestri D (2021) Scienza del Disegno. Città studi editore

Docci M (1985) Manuale di disegno architettonico, Bari

Filosto R, Inzerillo M (1974) Contributo dell'omologia nella rappresentazione grafica, Palermo. Quaderni di Dipartimento

Filosto R (1968) Lineamenti teorici del disegno, Palermo

De Fiore G (1964) Teoria della rappresentazione architettonica, Facoltà di Architettura di Roma. Quaderni di Dipartimento

De Fiore G (1967) Dizionario del disegno, Brescia

Fumoto K (1996) The eaves camber design method of Japanese traditional architecture by CAD. J. of architecture. Plan Environ Eng Trans AIJ 490(Dec. 1996):155–162

Glaeser G, Stachel H, Odehnal B (2016) The universe of conics. From the ancient Greeks to 21st century developments. Springer Spektrum, Berlin, Heidelberg. ISBN 978-3-662-45449-7. https://doi.org/10.1007/978-3-662-45450-3

L. Inzerillo, *The New Horizons of Oblique Axonometry*, UNIPA Springer Series, https://doi.org/10.1007/978-3-032-07475-1

Hofer M, Odehnal B, Pottmann H, Steiner T, Wallner J (2005) 3D shape recognition and reconstruction based on line element geometry. In: 10th IEEE international conference on computer vision, vol 2. EEE Computer Society. ISBN 0-7695-2334-X

Honvault P (2023) Local geometry of polyhedra and Cauchy's rigidity theorem. J Geometry Graph 27(1):39–46

Hoschek J (1971) Liniengeometrie. Bibliographisches Institut, Zürich

Inzerillo M, D'Alessandro M, Pizzurro P (1983) Omografia e prospettiva, Facoltà di Ingegneria di Palermo, Quaderni di Dipartimento

Inzerillo M (1977) Lezioni di Applicazioni di Geometria Descrittiva, Palermo. Quaderni di Dipartimento

Inzerillo M (1977) Lezioni di Teoria del Disegno Prospettico, Palermo

Inzerillo M, D'Alessandro M (1978) Omologia e rette limiti nella prospettiva delle immagini riflesse, Palermo

Inzerillo L (2004) Procedure di costruzione della prospettiva nel progetto del disegno, in Disegnare idee e immagini, n. 28. Gangemi editore, Roma

Inzerillo L (2012) The ellipsoid in orthogonal axonometric: homology application. In: DISEGNARECON, vol 5, n° 9, pp 335–342. ISSN 18285961. Autorizzazioni del Tribunale di Bologna n. 7896 del 30/10/2008

Inzerillo L (2012) Assonometria e futuro. Aracne editrice

Inzerillo L (2012) Essere prospettici. Pensare e realizzare prospettive. Aracne editrice

Kiuchi O (2007) Gendai Touryou no Sekkeijutu [Design technique of contemporary master builders]. Shinkentiku-sya

Migliari R (2008) Prospettiva dinamica interattiva, (a cura di), edizioni Kappa, Roma

Migliari R (2008) Geometria descrittiva, vol 2. Città studi editore

Migliari R (2009) Geometria descrittiva. Metodi e costruzioni, vol 1. Città studi editore

Odehnal B, Stachel H, Glaeser G (2020) The universe of quadrics. Springer, Berlin, Heidelberg. https://doi.org/10.1007/978-3-662-61053-4

Odehnal B (2021) Isoptic ruled surfaces of developable surfaces. J Geometry Graph 25(1):1–17

Ouarab S, Chahdi AO, Izid M (2020) Ruled surface generated by a curve lying on a regular surface and its characterizations. J Geom Graph 24(2):257–267

Di Paola F, Inzerillo L (2012) Hyperboloid and parabolid in orthogonal axonometric. In: Proceeding in the 15th international conference on geometry and graphics, Montreal, Canada, 1–5 Agoust 2012. ISBN 978-0-7717-0717-9

Phta Nguyen N (2020) A generalization of desargue's involution theorem. J Geometry Graph 24(1):65–72

Phta Nguyen N (2021) Projective line revisited. J Geometry Graph 25(2):187–196

Pylipaka S, Nesvidomin A, Volina T, Trokhaniak A, Taras I (2024) Creating a ruled surfaces using the base curve of the Frenet trihedron, pp 73–86

De Rubertis R (1975) Geometria Descrittiva, Roma

Saccardi U (1977) Applicazioni di Geometria Descrittiva, Firenze

Sgrosso A (1996) La rappresentazione geometrica dell'architettura, applicazioni di geometria descrittiva, Napoli

De Simone M (1973) Percezione e disegno, Palermo. Quaderni di Dipartimento

Spadafora G (2022) Fondamenti e applicazioni di geometria descrittiva. Franco Angeli editori, ISBN 9788891788795

Togashi S (1997) Mokuzoukentiku Yanekouhou Sumituke zukai [Manual of roof design for wooden architecture]

Toulias T, Lefkaditis G (2017) Parallel projected sphere on a plane: a new planegeometric investigation. Int Electron J Geometry 10(1):58–80. https://doi.org/10.36890/iejg.584443

Unluturk Y, Cimdiker M, Ekici C (2016) Characteristic properties of the parallel ruled surfaces with Darboux frame in Euclidean 3-space. Commun Math Model Appl 1:26–43

Vagnetti F (1947) La prospettiva lineare, Roma

Veljan D (2021) Improved Euler's inequality in plane and space. J Geom 112:31. https://doi.org/10.1007/s00022-021-00595-2

Wieleitner H (1908) Spezielle Ebene Kurven. Sammlung Schubert LVI, G. J. Göschen'sche Verlagshandlung, Leipzig

Wunderlich W (1937) Die isoptischen Kurven der Zykloiden. ZAMM Z Angew Math Mech 17(1):56. https://doi.org/10.1002/zamm.19370170108

Wunderlich W (1962) Autoevoluten. Elem Math XVII(6)

Wunderlich W (1971) Kurven mit isoptischem Kreis. Aequationes Math 6(1):71–81. https://doi.org/10.1007/bf01833240

Wunderlich W (1972) Kurven mit isoptischer Ellipse. Monatsh Math 75(4):346–362. https://doi.org/10.1007/bf01303569

MIX
Papier aus verantwortungsvollen Quellen
Paper from responsible sources
FSC® C105338

If you have any concerns about our products,
you can contact us on
ProductSafety@springernature.com

In case Publisher is established outside the EU,
the EU authorized representative is:
Springer Nature Customer Service Center GmbH
Europaplatz 3, 69115 Heidelberg, Germany

Printed by Libri Plureos GmbH
in Hamburg, Germany